T0136857

Sustainable Civil Infrastructures

Sustainable Infrastructure impacts our well-being and day-to-day lives. The infrastructures we are building today will shape our lives tomorrow. The complex and diverse nature of the impacts due to weather extremes on transportation and civil infrastructures can be seen in our roadways, bridges, and buildings. Extreme summer temperatures, droughts, flash floods, and rising numbers of freeze-thaw cycles pose challenges for civil infrastructure and can endanger public safety. We constantly hear how civil infrastructures need constant attention, preservation, and upgrading. Such improvements and developments would obviously benefit from our desired book series that provide sustainable engineering materials and designs. The economic impact is huge and much research has been conducted worldwide. The future holds many opportunities, not only for researchers in a given country, but also for the worldwide field engineers who apply and implement these technologies. We believe that no approach can succeed if it does not unite the efforts of various engineering disciplines from all over the world under one umbrella to offer a beacon of modern solutions to the global infrastructure. Experts from the various engineering disciplines around the globe will participate in this series, including: Geotechnical, Geological, Geoscience, Petroleum, Structural, Transportation, Bridge, Infrastructure, Energy, Architectural, Chemical and Materials, and other related Engineering disciplines.

More information about this series at http://www.springer.com/series/15140

Mohamed Shehata · George Anastasopoulos ·
Mattei Norma
Editors

Recent Technologies in Sustainable Materials Engineering

Proceedings of the 3rd GeoMEast International Congress and Exhibition, Egypt 2019 on Sustainable Civil Infrastructures – The Official International Congress of the Soil-Structure Interaction Group in Egypt (SSIGE)

Springer

Editors
Mohamed Shehata
Expertise House
for Engineering Consultations
Cairo, Egypt

George Anastasopoulos
International Accreditation Service (IAS)
California, CA, USA

Mattei Norma
University of New Orleans
New Orleans, LA, USA

ISSN 2366-3405 ISSN 2366-3413 (electronic)
Sustainable Civil Infrastructures
ISBN 978-3-030-34248-7 ISBN 978-3-030-34249-4 (eBook)
https://doi.org/10.1007/978-3-030-34249-4

This Springer imprint is published by the registered company Springer Nature Switzerland AG
The registered company address is: Gewerbestrasse 11, 6330 Cham, Switzerland

Contents

About the Editors

Dr. Mohamed Shehata, PhD, MBA, MSC, PMP, CLAQ, ASQ, CEO, is Founder of the EHE Consulting Group in Middle East.

He has more than 25 years' experience of many mega, large, and small projects in the Middle East. He was Leader of the multidisciplinary engineering works, so he has gained experience in the architectural, master planning, urban planning, project management, project preparations, decision making, and value engineering of the projects. In addition to all previous manager works, he has a professional expertise in the geotechnical, structural, and bridge engineering.

Dr. George Anastasopoulos is Vice President of International Accreditation Service (IAS). He is Mechanical Engineer with an MSc and a PhD in Applied Mechanics from Northwestern University. He is Member of ISO TC176 and CASCO committees who developed the new ISO 9001 and ISO/IEC 17025. He is awarded with the EOQ Presidential Georges Borel Award for international achievements in quality. He presented lectures related to Accreditation, Conformity Assessment, Management Systems, BPR, Telecoms-FTTH-IT and Process Auditing. He participated in consulting and research projects in USA, European Union, and many other countries worldwide.

Norma Jean Mattei, Ph.D., P.E., is Professor and Past Chair at the University of New Orleans' (UNO) Department of Civil and Environmental Engineering. She has been active in ASCE for more than 20 years in local, regional, and national leadership roles and was elected by the Society's membership as the 2017 ASCE President.

She sits on the executive committee of ASCE's New Orleans Branch SEI/Structures Committee. The former region 5 director has served on ASCE's Committee on Diversity and Women in Civil Engineering and the Committee on Licensure and Ethics. ASCE has drawn on her expertise for a number of media relation activities, including an interview with National Public Radio's "Morning Edition" on post-Hurricane Katrina conditions. Recently, she was spokeswoman for "Raised Floor Living," a commercial that aired in the New Orleans region promoting the elevation of residential structures above the floodplain.

In 2012, President Obama named Mattei one of three civilian members of the Mississippi River Commission, which researches and provides policy and work recommendations covering flood control, navigation, and environmental projects. In that capacity, she helped oversee a drainage basin that covers 41 percent of the nation. The governor of Louisiana appointed her to the state's licensing board for professional engineers, LAPELS. She also serves on the board of directors for both the Louisiana Transportation Research Center Foundation and the Louisiana Technology Council.

She has been Member of the UNO faculty since 1995. Her technical research interests include large watershed management, material and structural testing, sustainable reuse of spent construction and fabrication materials, and residual stress measurement. She is also interested in diversity, licensure, and ethics issues.

She earned a bachelor's degree in civil engineering in 1982 and a doctorate in 1994, both from Tulane University.

Development and Performance of Manual Technique Used in Production of Compressed Earth Blocks

Mohamed Darwish[1,2](✉), Safwan Khedr[3], Fady Halim[4], and Rana Khalil[4]

[1] Ahram Canadian University, Giza, Egypt
mdarwish@aucegypt.edu
[2] Adjunct Faculty, American University in Cairo, Cairo, Egypt
[3] American University in Cairo, Cairo, Egypt
[4] American University in Cairo, Cairo, Egypt

Abstract. Earth block construction is a low energy alternative when compared to conventional building materials, it is cost-effective, eco-friendly and safe. However, the technique of compressing such blocks have been always challenging as it has been typically an automated or semi-automated process that involves high costs. This study is part of a broad research at the American University in Cairo on earth construction materials. The goal of this paper is to develop and assess the performance of in-house designed manual equipment capable of compressing earth blocks that satisfy the strength and construction requirements. First, the equipment was designed and manufactured using locally-available low-cost components that could be readily available for the general public. The performance of this equipment was assessed in terms of studying the compressive strength of the manufactured compressed earth blocks (CEBs) and comparing them to their mechanically manufactured counterparts. Furthermore, the effects of changing the mix designs and varying the block thickness were studied. The performance was further assessed by studying the time taken by a typical user to use the equipment to manufacture compressed earth blocks and assess the productivity of this equipment.

1 Introduction

The unique traditional way of construction using earth blocks in hot-climate countries was a solid part of the cultural heritage existing for millennia due to its cost-effectiveness and eco-friendliness. This traditional building construction was very simple in terms of using the locally available earth as a raw material. However, it didn't produce buildings durable enough to withstand high loads, rain or even high humidity (Shetawy and Abdel-Latif 2008). Hence, modernity started to come into figure in the form of replacing the traditional earth blocks with reinforced concrete (RC) which is stronger and more durable than the traditional building construction; but more expensive, less eco-friendly in addition to replacing the cultural heritage represented in the culturally rich architectural style (Zhong and Wu 2015). Hence, the application of new construction technologies to preserve the culturally rich architecture through

M. Shehata et al. (Eds.): GeoMEast 2019, SUCI, pp. 1–12, 2020.
https://doi.org/10.1007/978-3-030-34249-4_1

providing environmentally and economically sustainable innovative designs of buildings would solve the major problems related to RC construction. One of these technologies is earth-based construction and specifically compressed earth blocks (CEBs).

Several researchers have studied the manufacturing and properties of CEBs. Some of these researchers like (Morel et al. 2007) added cement to the earth mixture used in producing compressed earth blocks in order to increase the block compressive strength and durability however this defeats the purpose of making compressed earth blocks from the first place as one of its major merits is its eco-friendliness and the fact that all of its raw materials are either natural or industrial by-products. Other researchers like (Agha 2003) and (Khedr et al. 2003) used lime as a stabilizer instead of cement which is much eco-friendlier and still achieving satisfactory levels of compressive strength.

On the other hand, some researchers started adding other natural materials to enhance the properties of CEBs. One of these studies was performed by (Taallah et al. 2014) who added palm fibers to the earth mixture and studied the effects of varying the percentage of palm fibers and the percentage of cement added within the mixture on the block compressive strength. The highest strength they attained was achieved by a block having 8% cement, 0.05% fibers and molded under a pressure of 10 MPa.

Meanwhile, other researchers like (Donkor and Obonyo 2015) have assessed the effect of adding polypropylene fibers to the mixture on the compressive strength and ductility of the produced CEBs. However, the fact that these fibers are expensive (when compared to earth or palm fibers) while they didn't make a major breakthrough in the properties of the blocks make using them not the best option from a cost-benefit perspective.

Furthermore, the area of thermal properties in relation to the density and compressive strength of CEBs was studied by (Ben Mansour et al. 2016). Within that study it was found that although the compressive strength was found to be in a direct relationship with the CEB density the thermal conductivity was found to have an inverse relationship with the CEB density. Hence, these researchers concluded that in order to satisfy both strength and thermal insulation criteria a density within the range of 1750 kg/m^3 should be targeted.

Similarly, (Sitton et al. 2018) studied the variation of mix proportions on the compressive strength of the CEBs however these CEBs were molded mechanically at a high pressure of 15.5 MPa that is manually unachievable in addition to the fact that they contained 11% of cement which carries the same problem as the work performed by (Morel et al. 2007) and (Taallah et al. 2014).

This need to get cement out of the mixture without sacrificing strength has initiated the research performed by (Sore et al. 2018) who used a synthesized geo-polymer made from natural raw-materials instead of cement within the mixture. The produced blocks were tested for both strength and thermal conductivity and was find to achieve satisfactory results when comparing them to those of CEBs involving cement within their mixtures. However, the economic side of using this geo-polymer was not studied by these researchers and the costs of its production was not mentioned within that study which raises a question mark on the feasibility of using such a synthesized material.

Although the efforts performed by the aforementioned researchers and other researchers to explore the mechanical and thermal properties of compressed earth blocks are significant, no researcher so far has developed a simplified manual method to compress earth blocks in an efficient and simple manner.

Consequently, the authors of this paper have put extensive efforts into developing a technique involving manually operated equipment that could efficiently produce CEBs with sufficient strength. The efficiency of the equipment was evaluated by comparing the strengths and manufacturing times of the CEBs produced by it to those produced by fully-mechanized equipment. The proper mix design was chosen based on a study of the compressive strength of CEBs having three different mixtures. The proper thickness of the CEBs was determined based on a study of the compressive strength and unit weight of CEBs having different thicknesses using the developed equipment.

2 Equipment Design and Manufacturing

In order to meet the needs of the general public who need a simple and economic way to manufacture CEBs, a setting was designed to be able to apply the molding compression force manually. The manual equipment consisted of the following main components:

1. A manual hydraulic piston to apply a vertical load of 320000 N on the top of the mixture. As shown in Fig. 1, the piston base is placed on top of a steel block while the piston itself will exert the force at the mid-point of the upper W310x52 I-beam.

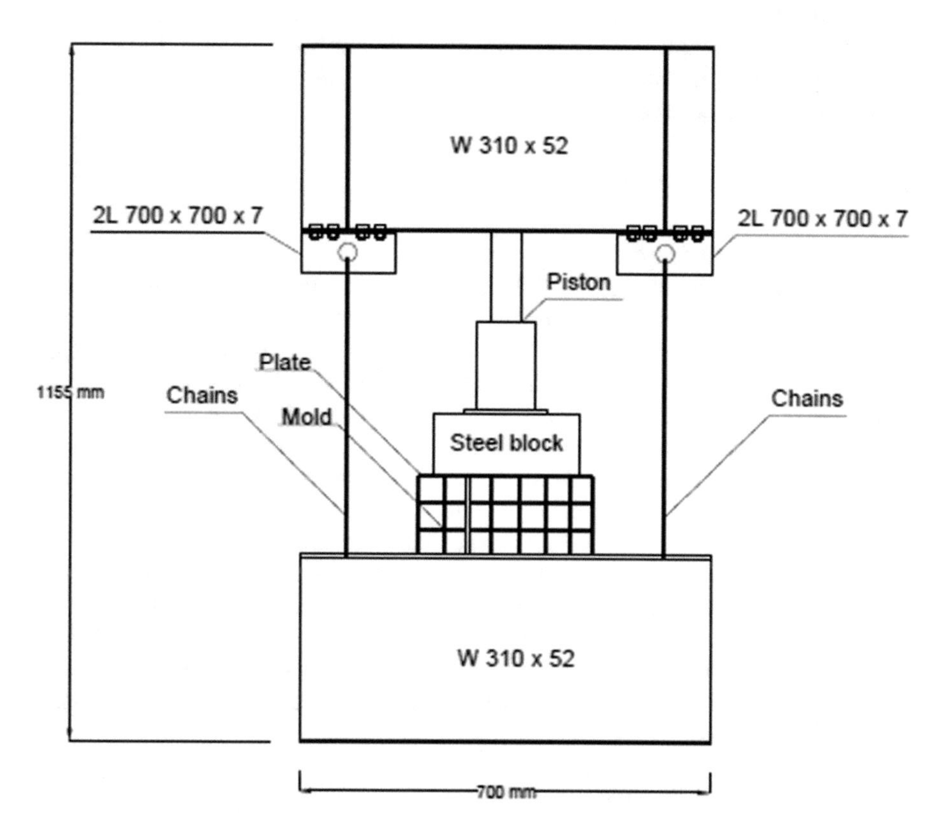

Fig. 1. Manually operated compressing equipment set-up.

2. A 100 mm high, 250 mm long and 100 mm wide steel block on which the base of the hydraulic piston rests as shown in Fig. 1.
3. A 5 mm-thick plate having an area of 300 mm × 140 mm that transfers the force from the steel block to the soil mixture within the mold.
4. A mold having an inner dimension of 300 mm long, 140 mm wide and 130 mm depth containing the soil mixture to be compressed.
5. Two W310x52 wide flange I-beams; one placed beneath the mold while the other is placed above the hydraulic piston as shown in Fig. 1. The top beam is stiffened by stiffeners at the location of the connection with the double angles as shown in Fig. 1.
6. Two pairs of double angles (back-to-back) connected to the lower flange of the upper beam as shown in Fig. 1.
7. Four chains connecting the lower beam to the double angles as shown in Fig. 1. The chains will remain in tension as long as the force is applied and the compression process is in action.

The beams were both designed to carry bending moments and shear forces caused by the 320000 N concentrated force. Each chain was checked to be able of carrying a capacity of more than the 80000 N that each will carry. The bolted connections between the upper I-beam and the double angles were designed. The connection between the chains and the double angles was also checked for bearing and shear. Two molds were manufactured having an inner dimension of 300 mm long, 140 mm wide and 130 mm deep in order to produce blocks that are 300 mm × 140 mm with thicknesses that could reach up to 130 mm.

2.1 Equipment Efficiency Assessment

Block Strength Assessment

In order to assess the efficiency of the developed manually operated equipment in effectively molding and compressing the earth blocks, three blocks were manufactured using 46% sand, 46% loam and 8% lime as developed by (Agha 2003) and (Khedr et al. 2003) and compressed till reaching a thickness of 90 mm using the compressing testing machine. Three other blocks having the same dimensions and incorporating the same mix design have been produced using the manually operated equipment. The two groups of blocks were tested under compression till failure at the age of seven days. The stress-strain diagrams presented in Fig. 2 show that the ultimate compressive strengths of the blocks manufactured using the manually operated equipment fall within the same range of values as those achieved by the mechanically manufactured blocks. Furthermore, Fig. 3 shows that the moduli of toughness of the manually manufactured blocks fall within the same range of their mechanically manufactured counterparts with minor variations. These results confirm that the developed manually operated equipment could compress earth blocks as efficiently as the mechanical compression machine.

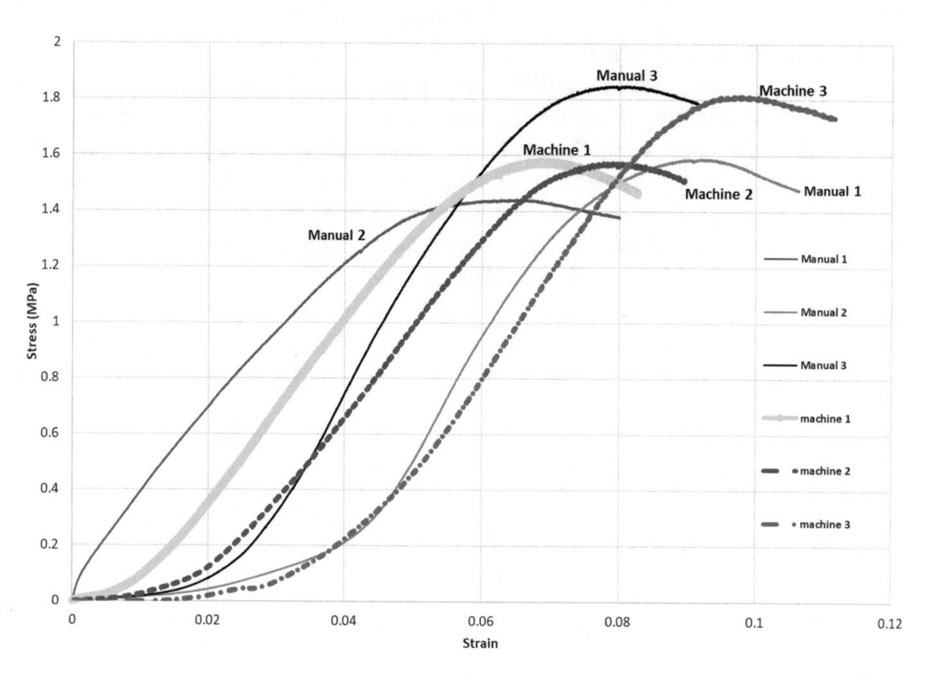

Fig. 2. The stress-strain diagrams for the blocks manufactured mechanically versus their manually manufactured counterparts.

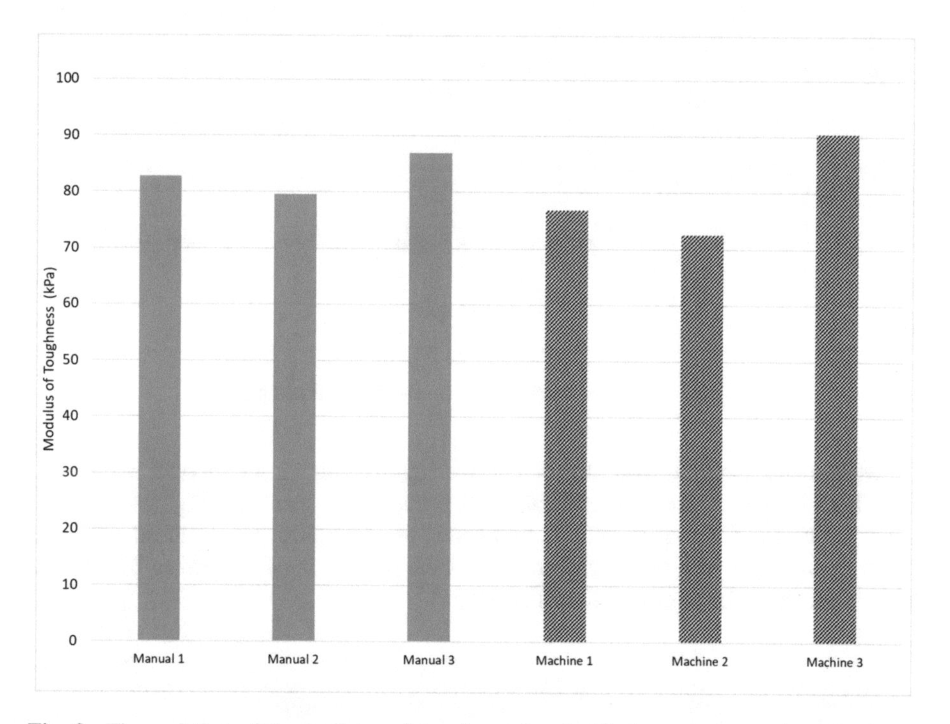

Fig. 3. The variation of the modulus of toughness for the blocks manufactured mechanically versus their manually manufactured counterparts.

Compressing Time Assessment
In order to assess the efficiency of the developed manually operated equipment in molding and compressing the blocks in a time-efficient manner, the time of manufacturing nine blocks done within this study was monitored. This time included filling the mold with the mixture, compressing the mixture to form the block and unpacking the block from the mold. The results are shown in Fig. 4 in which the manufacturing time is plotted versus the sample number. This curve represents the learning curve for a typical general labor who could efficiently manufacture a block within seven minutes and a half after using the equipment for the ninth time. Hence, theoretically speaking, a labor could produce six to eight blocks within an hour. However, taking the efficiency of labors into consideration, this number will practically drop to be four blocks per hour.

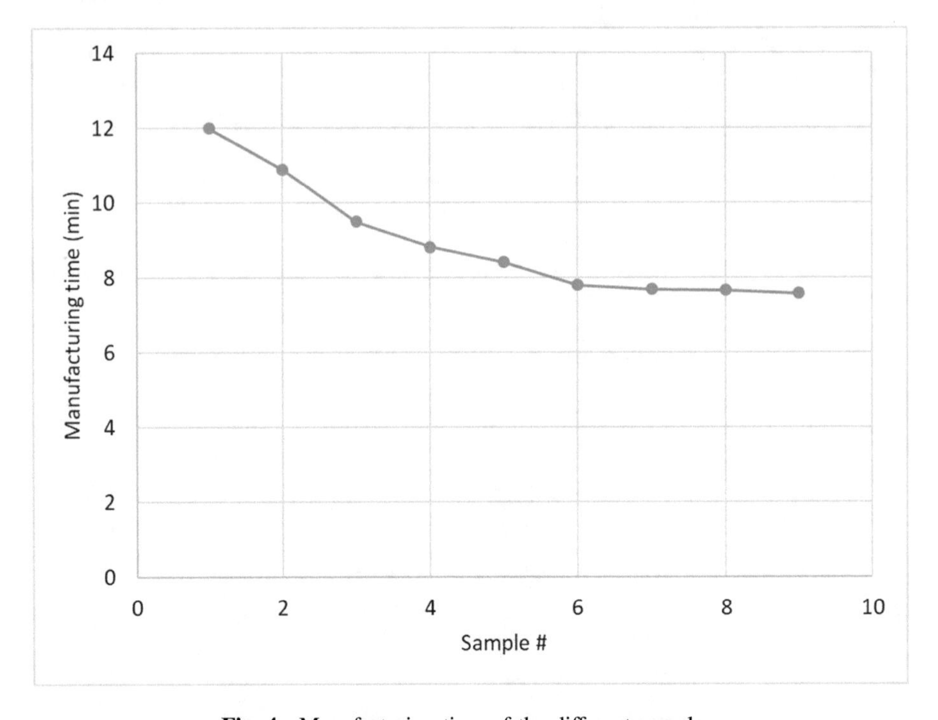

Fig. 4. Manufacturing time of the different samples.

Meanwhile, the processes of grinding, sieving and mixing the mixture ingredients have been found to take approximately 3–4 h due to the loam and lime coming in a solidified form. Consequently, ten to twelve blocks could be produced within a working day of eight hours. This proves that the process is considered to be efficient and simple even for a general unskilled labor.

3 Block Mix Design and Dimensions

3.1 CEB Mixture Choice

Three mix designs for the CEBs have been prepared and tested, two of them were the mixtures previously produced by (Agha 2003) and (Khedr et al. 2003) involving lime and palm fibers separately. Hence it was necessary to study a third mix that is actually a combination of the first two involving both lime and palm fibers. All of the mixtures contain loam and sand as these are the materials available most in the western desert of Egypt. For all of the mixtures the ratio between sand and loam is kept as 1:1 as this is the same ratio used by (Agha 2003) and (Khedr et al. 2003) while the percentages of lime and palm fibers varied as shown in Table 1. Meanwhile the molding pressure was 7.6 MPa as it depended on the force exerted by the hydraulic piston in the manual equipment for all manufactured blocks.

Table 1. Block Mix designs

Mix #	Constituents			
	Loam	Sand	Lime	Palm fibers
I	46%	46%	8%	0
II	49.5%	49.5%	0	1%
III	46%	46%	7%	1%

Three blocks of each mix were produced and tested at the ages of 7 days for compressive strength. Depending on the results of the 7-day strength tests the best mix is chosen and will be used to construct the prisms subjected to compression and walls subjected to lateral loads in further research.

The three mixtures having different percentages of lime and palm fibers were tested under compression after seven days. Three blocks of each mixture were subjected to compression till failure. During each test, the deflection and load was recorded and used to calculate the values of stresses and strains. The stress-strain diagrams for the nine tested samples are shown in Fig. 5. The overall ultimate strength of the mixtures involving 8% lime and no palm fibers is higher than that of the other two mixtures. Meanwhile it could be noticed that the plastic strains in the samples involving palm fibers (whether with or without lime) were generally higher than the plastic strains in the samples having no palm fibers. This indicates that although including palm fibers in the mixture decreases its ultimate strength, these fibers increase the ductility of the blocks. This could be obviously recognized in the moduli of toughness presented in Fig. 6 as the moduli of toughness of the blocks involving both palm fibers and lime had the highest moduli of toughness due to the fact that they have experience higher strains before failure.

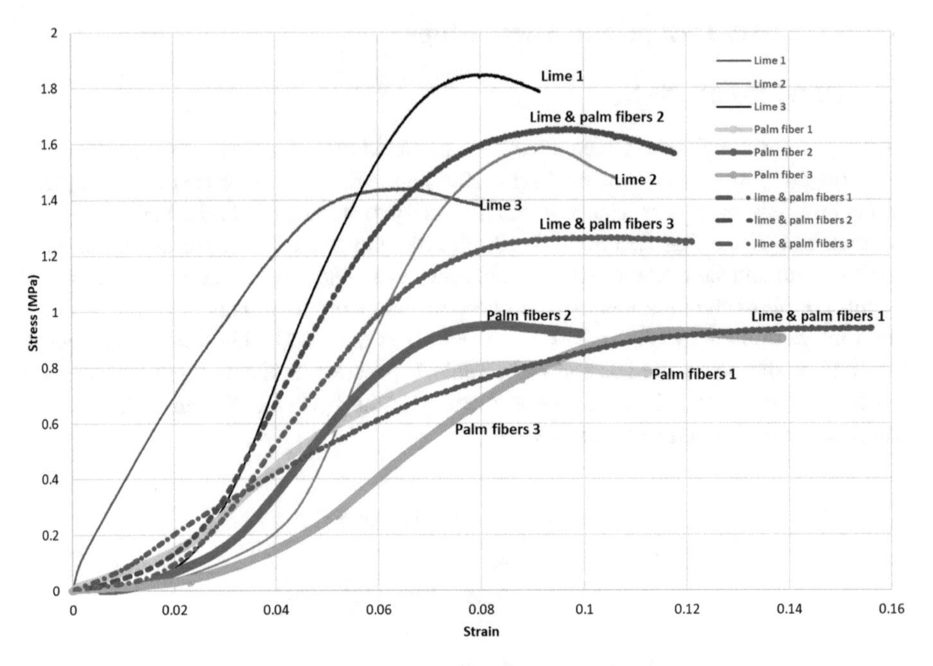

Fig. 5. Stress-strain diagrams for the three different mixtures.

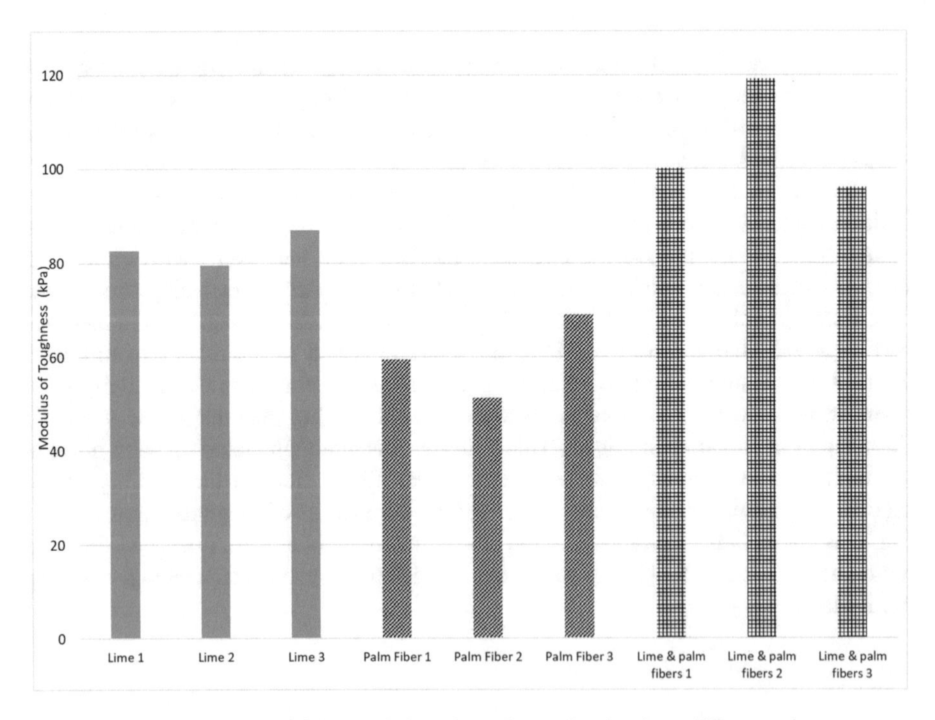

Fig. 6. The variation of the modulus of toughness for the three different mixtures.

On the other hand, (New Zealand Standards 1998) state that the minimum allowable compressive strength is 1.3 MPa. Hence, and since the compressive strength within practical applications is more important than ductility and since the ultimate strength of the samples having 8% lime and no fibers ranges between 1.44 MPa and 1.85 MPa with an average value of 1.68 MPa, the mixture involving 8% lime and no fibers is the one selected to be used for the remaining part of the study.

3.2 CEB Thickness Choice

Furthermore, the mix having the highest compressive strength was used to compress mixtures with three different degrees of compression achieving three different thicknesses which were 110 mm, 100 mm and 90 mm and comparing the compressive strength of the three produced.

In order to assess the effectiveness of the compaction process using the newly manufactured manual equipment three groups of samples having three different thicknesses (90 mm, 100 mm and 110 mm) were manufactured and tested under compression after seven days. During manufacturing, each mold was filled to its full depth of 130 mm however during applying the molding compression each group of samples was molded and compressed until reaching its specified thickness. At the age of seven days, three blocks of each mixture were subjected to compression till failure. During each test, the deflection and load were both recorded and used to calculate the values of stresses and strains. The stress-strain diagrams for the nine tested samples are shown in Fig. 7. The 90 mm thick samples were found to have the highest strength when compared to the thicker samples. This could be also noticed when comparing the area under each of the stress-strain diagrams which represents the toughness as the toughness of the 90 mm thick samples is obviously higher than that of the other samples. This could be obviously recognized in the moduli of toughness presented in Fig. 8 as the moduli of toughness of the blocks have increased with the decrease in thickness as the 90 mm thick samples had the highest moduli of toughness due to the fact that they have experience higher strains before failure.

The increase in strength and toughness with the increase in compaction is expected however what was unexpected was the degree of that increase in strength that was nearly doubled. This could be linked to a fact that the researchers involved in the manual compaction process have noticed during compressing the 90 mm samples which is that the handle of the manual hydraulic piston becomes significantly heavy and needs significant effort to operate when compressing the last 10 mm (i.e. the difference between the 100 mm and the 90 mm thicknesses).

On the other hand, the variation of the unit weight was studied. As shown in Table 2, the unit weight increases with the decrease in thickness reflecting a higher degree of compaction however the difference in unit weight between the 90 mm thick group and the 100 mm thick group is less than the difference in unit weight between the 100 mm thick group and the 110 mm thick group. This also explains why it was extremely difficult to compress the blocks to reach a thickness less than 90 mm manually which would have been expected to be of higher compressive strength however, this was not performed as the current research targets enabling the average person to be able of performing the block manufacturing process manually.

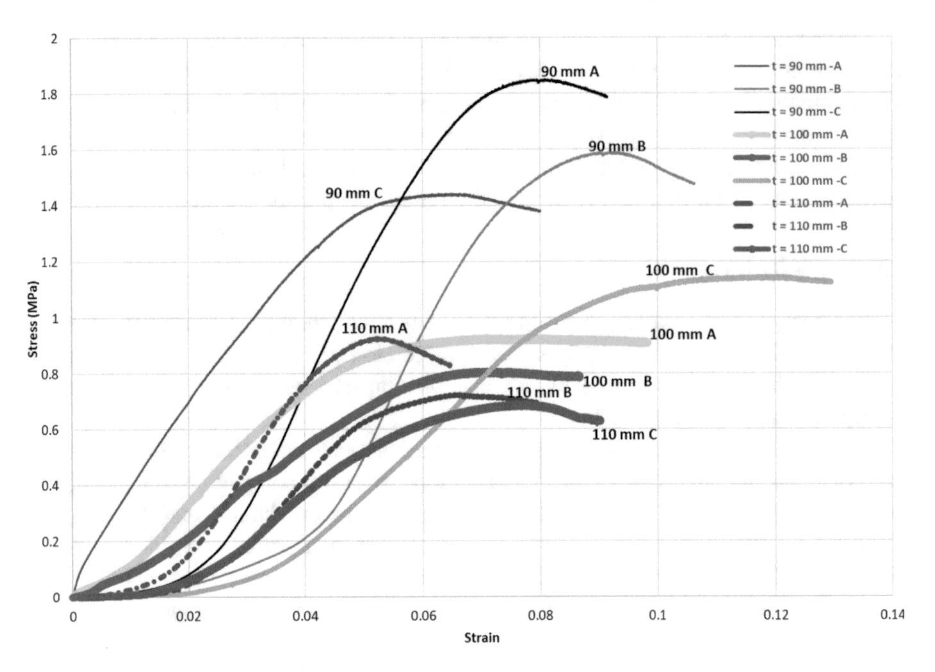

Fig. 7. Stress-strain diagrams for blocks with different thicknesses.

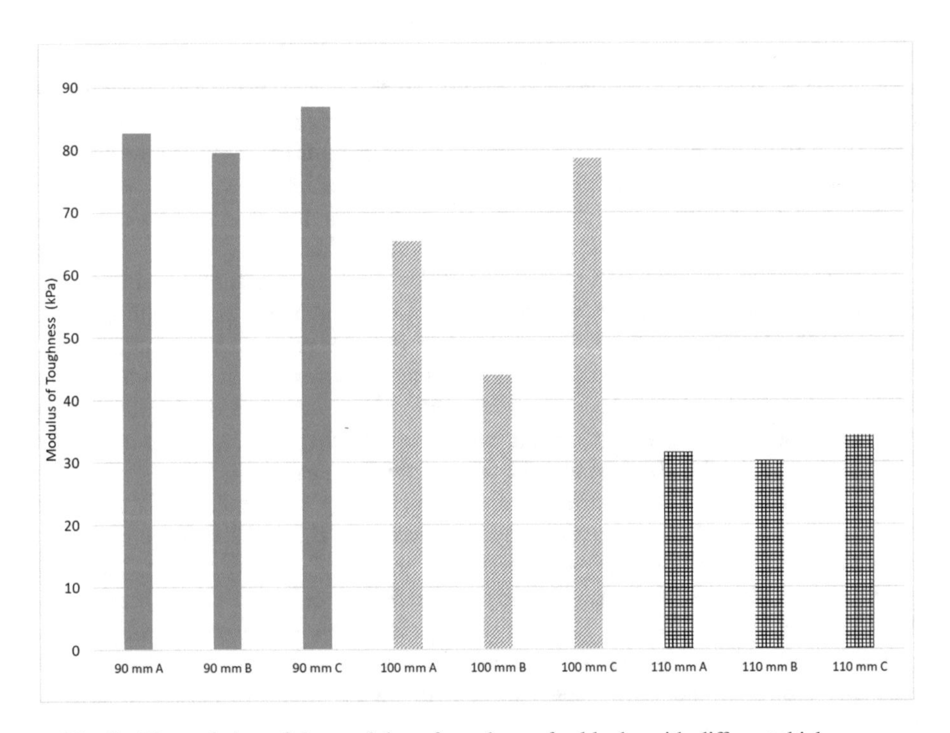

Fig. 8. The variation of the modulus of toughness for blocks with different thicknesses.

Table 2. Unit weights of the blocks having different thicknesses.

Group	Sample	Weight (kg)	Volume (m^3)	Unit weight (kg/m^3)	Average unit weight (kg/m^3)
90 mm	90 mm A	7.98	0.00378	2111	2058
	90 mm B	7.6	0.00378	2011	
	90 mm C	7.76	0.00378	2053	
100 mm	100 mm A	8.4	0.0042	2000	2018
	100 mm B	8.21	0.0042	1955	
	100 mm C	8.815	0.0042	2099	
110 mm	110 mm A	8.45	0.00462	1829	1864
	110 mm B	8.715	0.00462	1886	
	110 mm C	8.675	0.00462	1878	

4 Conclusions

Based on the results reached and analysis performed the following conclusions could be drawn:

- The developed manually operated equipment is as efficient in molding as mechanical compressive machines.
- The developed manual technique is time-efficient enabling manufacturing of 10–12 blocks per day.
- The best mix for block production is the one containing 46% sand, 46% Loam and 8% lime.
- Introducing palm fibers increases ductility but decreases the strength.
- The strength is inversely related to the block thickness.
- The unit weight is inversely related to the block thickness.
- The manually achievable thickness of the strongest block produced is 90 mm.

Acknowledgments. The authors would like to acknowledge the funding received from the American University in Cairo. The authors would also like to acknowledge the efforts of the lab personnel in the Construction Engineering Department at the American University in Cairo.

References

Agha, S.: Performance of compressed earth blocks. Master's thesis. Cairo, Cairo. The American University in Cairo, Egypt (2003)

Ben Mansour, M., Jelidi, A., Cherif, A.S., Jabrallah, S.B.: Optimizing thermal and mechanical performance of compressed earth block (CEB). Constr. Build. Mater. **104**, 44–51 (2016). https://doi.org/10.1016/j.conbuildmat.2015.12.024

Donkor, P., Obonyo, E.: Earthen construction materials: assessing the feasibility of improving strength and deformability of compressed earth blocks using polypropylene fibers. Mater. Des. **83**, 813–819 (2015). https://doi.org/10.1016/j.matdes.2015.06.017

Khedr, S., Abou-Zeid, M.N., Agha, S.: Concerns in manufacturing compressed earth blocks. In: Proceedings of the 35th Canadian Society of Civil Engineering Annual General Conference. Moncton, Canada: CSCE (2003)

Morel, J.-C., Pkla, A., Walker, P.: Compressive strength testing of compressed earth blocks. Constr. Build. Mater. **21**, 303–309 (2007). https://doi.org/10.1016/j.conbuildmat.2005.08.021

New Zealand Standards. NZS 4298: Materials and workmanship for earth buildings. NZS 4298: 1998. Standards New Zealand, Wellington, New Zealand (1998)

Shetawy, A.A., Abdel-Latif, M.M.: Echoes of the environment: housing patterns in Siwa Oasis Egypt. Ain Shams J. Archit. Eng. **118**, 17–29 (2008). https://doi.org/10.2495/STR110331

Sitton, J.D., Zeinali, Y., Heidarian, W.H., Story, B.A.: Effect of mix design on compressed earth block strength. Constr. Build. Mater. **158**, 124–131 (2018). https://doi.org/10.1016/j.conbuildmat.2017.10.005

Sore, S.O., Messan, A., Prud'homme, E., Escadeillas, G., Tsobnang, F.: Stabilization of compressed earth blocks (CEBs) by geopolymer binder based on local materials from Burkina Faso. Constr. Build. Mater. **165**, 333–345 (2018). https://doi.org/10.1016/j.conbuildmat.2018.01.051

Taallah, B., Guettala, A., Guettala, S., Kriker, A.: Mechanical properties and hygroscopicity behavior of compressed earth block filled by date palm fibers. Constr. Build. Mater. **59**, 161–168 (2014). https://doi.org/10.1016/j.conbuildmat.2014.02.058

Zhong, Y., Wu, P.: Economic sustainability, environmental sustainability and constructability indicators related to concrete- and steel-projects. J. Clean. Prod. **108**, 748–756 (2015). https://doi.org/10.1016/j.jclepro.2015.05.095

Efficacy of Treatments on Coal Bottom Ash as a Cement Replacement

Sharon Gooi, Ahmad A. Mousa(✉), and Daniel Kong

Monash University, Clayton, Malaysia
ahmad.mousa@monash.edu

Abstract. Coal bottom ash (CBA), a silica rich by-product of coal burning, has been primarily researched as a substitute construction material. This study investigates the effect of the treatment type and optimum dosage of CBA as a binder replacement in cement paste. Pulverisation, soaking and burning of CBA were conducted prior to mixing in order to enhance its innate physical and chemical characteristics. Materials from two Malaysian power plants, Kapar (K) and Tanjung Bin (TB), were used to examine the 'source effect' on the pozzolanic constituents of CBA. Subsequently, the compressive strength of the produced cement paste was investigated. The total pozzolanic content (SiO_2, Al_2O_3, and Fe_2O_3) in CBA from K amounts to 63%, whereas it is merely 50% for TB. The compressive strengths of cement paste involving CBA at different levels of binder replacement were compared to gauge the effect of the treatment type. The cement paste made with untreated (raw) TB and K-CBA have both shown comparable and continuous declining trends in the compressive strength up to 50% cement replacement. At a liquid-to-binder (l/b) ratio of 0.35, the raw CBA samples showed relatively similar compressive strengths. For both sources, however, the compressive strengths of the pulverised CBA cement paste samples surpassed those utilising untreated CBA – signifying the positive impact of pulverising on the cementitious properties of CBA. The compressive strengths for the pulverised K-CBA and TB-CBA at 50% cement replacement were 24.3 and 27.3 MPa, respectively. These strengths are adequate for basic structural application, e.g. concrete pavers. Burning and soaking CBA have also improved the compressive strength, but to a lower extent.

1 Introduction

Coal fired power plants are one of the main sources of electricity worldwide. As technology advances, the efficiency of the coal boilers increases, allowing the use of different types of coal and reducing the energy loss of the process. Combustion of coal in industrial facilities produces two common by-products: fly ash (FA) and coal bottom ash (CBA). The amount of waste produced by burning coal is still substantial. The annual coal waste in the United States (American Coal Ash Association 2017) and Europe (ECOBA 2016) is estimated be on the order of 100 million and 40 million tons, respectively.

Considerable research has been conducted on the utilisation of FA in the construction industry as a Portland Cement (PC) replacement. FA particles are typically

M. Shehata et al. (Eds.): GeoMEast 2019, SUCI, pp. 13–21, 2020.
https://doi.org/10.1007/978-3-030-34249-4_2

spherical in shape and very fine, but CBA mostly consists of sand-sized angular particles. Unlike FA, CBA has yet to be adopted into the construction industry. Currently, CBA has no economic value beyond serving as a filler or fine aggregate replacement in concrete. This is in part due to its low pozzolanic reactivity in comparison to FA, resulting in low compressive strengths. Additionally, the appreciable variation in size and morphology of the CBA particles are deemed as unfavourable attributes. Better utilisation of CBA in construction necessitates enhancing its pozzolanic reactivity as part of the cementitious matrix. As such, CBA could be of a much higher economic value. This inevitably provides a more viable environmental solution in terms of reduced dependency on PC and river sand, and would equally eliminate the need for dumping CBA in landfills or ash ponds.

This study aims to investigate possible treatment methods to improve the pozzolanic reactivity of CBA such that it can be efficiently utilised as a cement replacement. Conducted research in the past five years, shows that the 12 out of 15 studies investigating CBA as PC replacement most commonly treated CBA through pulverising, otherwise known as grinding. Pulverising and sieving of the angular CB particles increase their surface area and subsequently enhances their reactivity. A treatment cycle consisting of drying, pulverising and sieving was performed for this purpose. The highest specific surface areas (SSA) of pulverised CBA recorded in literature (Argiz et al. 2017; Oruji *et al.* 2017) were 3,463 and 1,102 m^2/kg, respectively. These two studies achieved maximum compressive strengths of 68 and 31 MPa for mortars with CBA as 5 and 9% replacement of PC. It can be observed that CBA with higher SSA can produce better compressive strengths. Additives such as superplasticiser, lime or alkali activators are commonly used in tandem with pulverising, drying and sieving. Burning and soaking of CBA are, however, seldom carried out. These two treatments were conducted to reduce carbon content and other impurities or reduce water-trapping of CBA particles during mixing for the utilisation of CBA as a sand replacement (Jang *et al.* 2016). This study considered burning and soaking as a potential technique for removing impurities and increasing the reactivity of the CBA cement paste samples.

2 Material Preparation

CBA was obtained from two sources, the Tanjung Bin (Johor) and Kapar (Selangor) power plants in Malaysia. bituminous coal used in these plants has a wide range of carbon content of (45–86%), which is mostly burnt in the coal boilers (Morse and Turgeon 2012). The Blaine-based SSA of PC, pulverised TB-CBA and pulverised K-CBA were estimated to be 349.9, 133.4 and 217.7 m^2/kg, respectively. Due to the comparatively large particle size of the raw TB-CBA and K-CBA, the SSA could not be tested. CEM II class PC and tap water were used to make cement paste specimens.

The Los Angeles (LA) Abrasion machine was used to pulverise the raw (as received) CBA, using ten steel balls with a diameter of 47 mm. The CBA was placed in the LA machine drum and subjected to 2,000 to 10,000 revolutions during the grinding trials. Optimal grinding with minimal amount of wastage was found to be achieved at 4,000 revolutions. The pulverised CBA was removed from the drum and sieved.

Only the CBA that passed through a 150 μm sieve is utilised for mixing. The remaining material was returned to the LA Abrasion machine for further pulverising. Particle size distributions of the pulverised and sieved CBA are shown in Fig. 1. Burning of CBA was conducted in a muffle furnace at 700 °C for 2 h. After burning, changes in colour was observed for the CBA from both sources. TB-CBA turned from black to light grey, while K-CBA turned from grey to brown. This difference in colour is most likely due to the removal of unburnt carbon and other impurities. Soaking was conducted by immersing the CBA in tap-water bath for 24 h. For both sources of CBA, a separation of particles after soaking was observed, with the denser particles settling at the bottom of the bath and the lighter (finer) particles suspended or floating on the surface. This phenomenon was observed more clearly in K-CBA due to its coarser gradation.

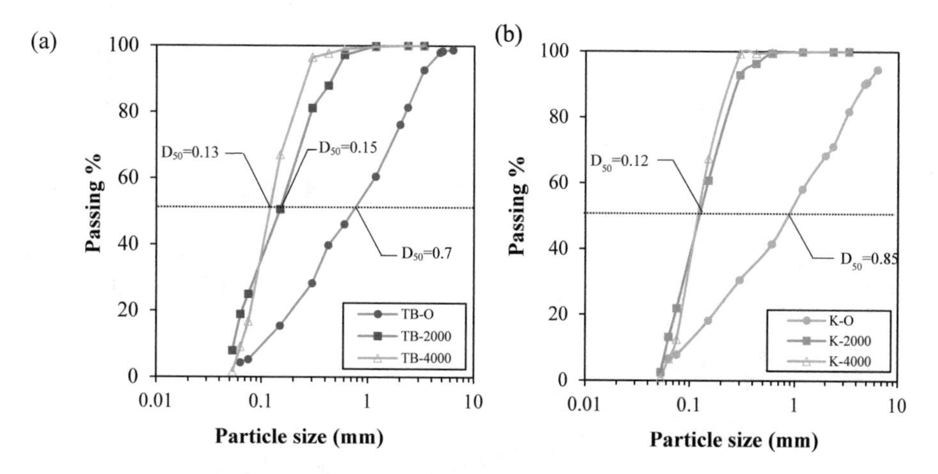

Fig. 1. Particle size distribution for original and pulverised CBA: (a) TB, (b) K.

3 Characterisation and Testing

Characterisation tests were conducted on the CBA from both sources. SSA was measured using a Blaine air-permeability apparatus (ASTM C204-18e1 2018). Chemical composition of the CBA was tested through X-ray fluorescence (XRF) and X-ray diffraction (XRD). The XRF results were compared with other studies that utilised CBA from the TB and K (Table 1). The XRD analysis for both sources is shown in Fig. 2. The K-CBA has 12.5% higher pozzolanic content (SiO_2, Al_2O_3, and Fe_2O_3) and 1.0% less loss on ignition (LOI) value than TB-CBA. In Table 1, it can be observed that the chemical composition of the CBA used in this study is significantly different from reported values in the literature for both facilities. This inconsistency in composition could be attributed to the continuous change in the coal source used in these power plants.

Table 1. Chemical composition of CBA from Tanjung Bin (TB) Kapar (K).

Content (wt%)	TB[a]	TB[b]	TB[c]	K[a]	K[d]	K[e]
SiO_2	29.5	45.3	46.6	44.0	56.0	68.9
Al_2O_3	10.8	18.1	23.6	14.9	26.7	18.7
Fe_2O_3	10.1	19.8	12.4	4.0	5.8	6.5
LOI	1.3	–	–	0.3	4.6	2.7

[a]This study
[b]Rafieizonooz *et al.* (2016)
[c]Latifi *et al.* (2015)
[d]Naganathan et al. (2015)
[e]Ibrahim *et al.* (2015)

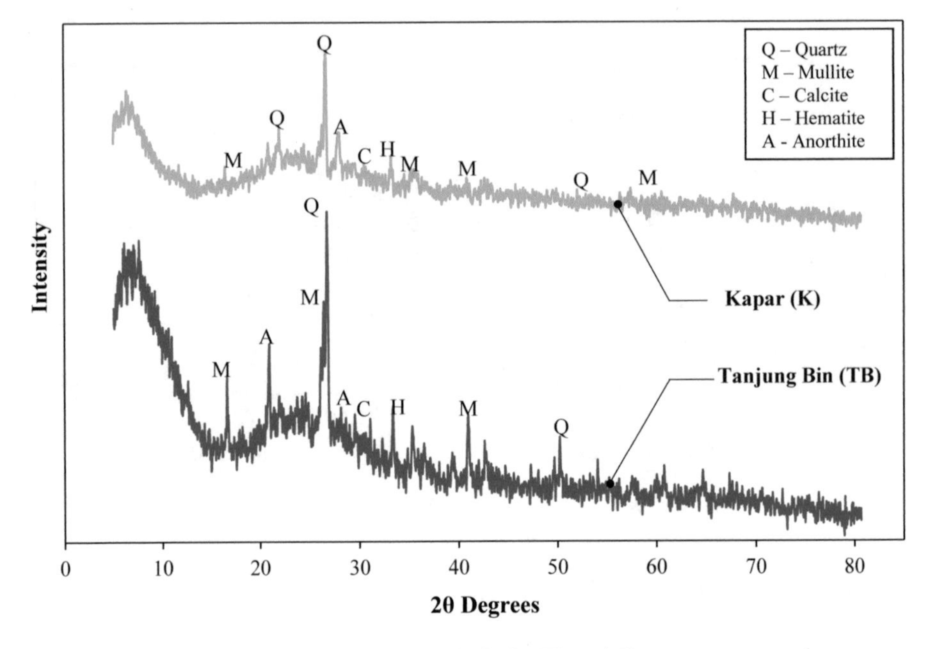

Fig. 2. XRD analysis for TB and K.

Triplicate cement paste cubes of 50 × 50 × 50 mm using CBA as a partial replacement of cement were cast using an l/b ratio of 0.35. This l/b ratio was chosen upon experimentation with l/b ratios ranging from 0.25 to 0.40. The 0.35 l/b ratio provides a good balance between strength and workability. This paper presents only the results for l/b of 0.35 mix. The raw and treated CBA was dry-blended prior to mixing with water. The cement paste samples were cast into the moulds in two layers of equal height. The moulds were placed on the vibrating table for 10 s to compact each layer. Thin polyethylene bags were placed over the moulds to prevent water evaporating and left for 24 h. The samples were subsequently removed from the moulds and water-cured for 28 days before conducting the compressive strength tests. The average compressive strength, standard deviation (0.61–10.67) and coefficient of variation (0.02–0.36) of the three samples in each mix were recorded. Control samples for both l/b ratios using the same Portland cement as in the CBA mixes were made using the same procedure. The notation used for the cementitious mixes was "source-treatment-replacement level", e.g. K-P-50 denotes pulverised Kapar CBA replacing 50% cement. Scanning electron microscopy (SEM) imaging was performed on the sides of the 28-day 50 × 50 × 50 mm cubes to observe their microstructure. Optical microscopy images of the samples were processed through MATLAB software (MATLAB R2017b 2017) to compare the porosity of the raw and treated CBA samples.

4 Results and Discussions

The control PC cement paste samples developed a 28-day compressive strengths of 50.8 MPa. Figure 3 depicts the 28-day compressive strength ($f_{cu,28d}$) for TB and K. Table 2 summarises the relative strength for all mixes with respect to the control mix, calculated as the ratio between the average compressive strength of each mix to that of the control mix (expressed as a percentage). The untreated CBA mixes performed relatively well at the 10–20% replacement levels with compressive strengths above 20 MPa. At 30–50% replacement, the compressive strength of the raw CBA samples decrease significantly and generally displays the lowest strengths of all the mixes (as low as 9.3 MPa).The untreated TB-CBA performs better than K-CBA, which can be attributed to the as received finer particle size distribution and subsequently the higher SSA of the TB-CBA.

The least effective treatment was soaking. It is likely that the water was unable to fully saturate the large, unbroken CBA particles, thus reducing the efficacy of this treatment. Most of the soaked CBA samples, yet, show higher compressive strength than the raw CBA mixes. There is better hydration reaction in the remaining PC content in the mixes due to the soaked CBA trapping less water. The second most effective

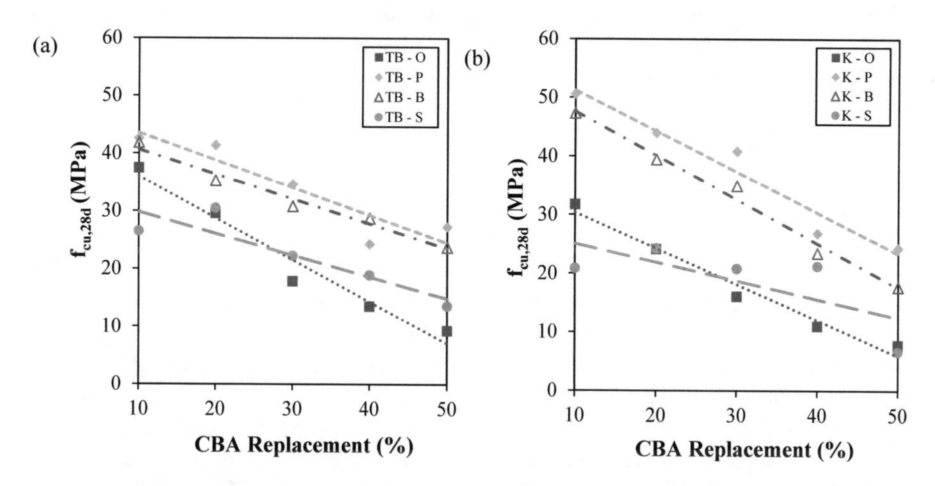

Fig. 3. Compressive strength of CBA cement paste: (a) TB; (b) K.

Table 2. 28-day compressive strength[a] of CBA specimens with respect to control specimen.

Replacement (%)	Tanjung Bin				Kapar			
	O	S	B	P	O	S	B	P
10	74	52[b]	82	84	65	41	93	100
20	58	60	69	82	56	48	78	87
30	35	44	61	68	37	41	69	80
40	27	37	57	48	27	42	46	53
50	18	27	47	54	18	13	35	48

[a]:Results shown are expressed as a percentage of control PC mix with 0% CBA
[b]:This result has a large coefficient of variation.
O–Original, S–Soaked, B–Burnt, P–Pulverised
Mixes will be denoted as Source-Treatment-Replacement Percentage (e.g. K-P-50 designates Kapar-Pulverised-50% replacement).

treatment is the burning of raw CBA. Carbon content is believed to reduce the pozzolanic reaction of the CBA and weaken the bonding with cement. The removal of the unburnt carbon and the impurities in the CBA after burning allowed for more complete PC hydration and enhanced the pozzolanic reaction of the CBA. As can be observed in Table 2 and Fig. 3, the burnt CBA specimens displayed compressive strength that is marginally less compared to the pulverised CBA, ranging from 17.7–47.3 MPa for both sources.

The pulverised CBA samples (K-P and TB-P) are consistently higher in compressive strength than the K-B samples. The strength of K-P-50 mix (24.3 MPa) is 13% higher than the that of K-B-50 mix (17.7 MPa), while the strength of TB-P-50 mix (27.3 MPa) is merely 7% higher than that of TB-B-50 sample (23.7 MPa). The cement paste samples made with pulverised CBA displayed the highest compressive strength of all the samples. The pulverised CBA had increased surface area compared to the other treated CBA, and this enhanced the pozzolanic reaction of the CBA, resulting in higher compressive strengths. At the highest CBA replacement (50%), the TB-P-CBA samples showed the largest compressive strength (27.3 MPa) as compared to all other treatments. This level of cement replacement is economically appealing. However, as with the untreated CBA mixes, an increase in the CBA content resulted in reduced compressive strength.

The other factors that affect the compressive strength are the source of the CBA and l/b ratio used for the mixes. The compressive strength of the CBA mixes largely depends on the composition of sources, more specifically the pozzolanic content. The K-CBA samples with pulverised or burnt CBA generally had higher compressive strengths compared to the TB-CBA, but most of the original and soaked TB-CBA samples have higher strength compared to the K-CBA counterparts. Considering the pozzolanic content and SSA of the CBA, these results are expected. The original TB-CBA is finer in size than the original K-CBA, and therefore has higher SSA, resulting in more complete pozzolanic reactions in the original TB-CBA mixes and better compressive strength. The soaked TB-CBA similarly has higher SSA than those for K-CBA, thus increasing the compressive strength.

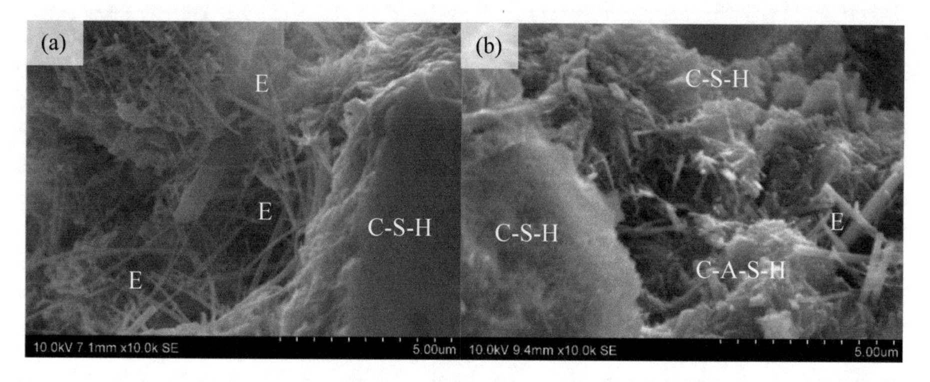

Fig. 4. SEM images for Kapar CBA: (a) K-O-50; (b) K-P-50.

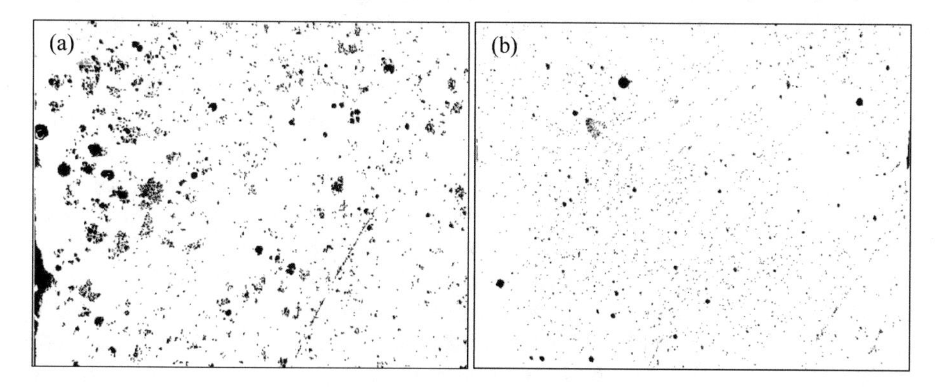

Fig. 5. Optical microscope images for Kapar CBA: (a) K-O-50; (b) K-P-50. (pores appear as black spots in the images - processed using MATLAB)

The microstructure of the 28-day cement paste specimens was examined using SEM imaging. The K-O-50 and the K-P-50 specimens were chosen due to the wide range in strength. The microstructure of these two mixes is presented in Fig. 4. The K-O-50 specimen had a large number of ettringite (E) needles throughout the microstructure of the sample (Fig. 4a). Some traces of calcium-silicate-hydrate (C-S-H) gel is observed, but not as extensive as those in the K-P-50 sample (Fig. 4b). Most of the E needles in the K-P-50 sample have destabilised to form monosulfate hydrate (C-A-S-H) crystals, and more C-S-H gel has formed. Therefore, it can concluded that the pozzolanic reactions in the K-P-50 sample were occurring at a much faster rate than that in the K-O-50 sample.

In addition to the noticeable differences in the microstructure, these two mixes have also portrayed differences in porosity. Figure 5 shows the optical microscope images of the K-O-50 and K-P-50 mixes that were post-processed using MATLAB software (MATLAB R2017b 2017b). The K-O-50 mix has a higher density of large pores compared to the K-P-50 mix. This is most likely due to the angular, relatively large size of the original CBA particles that allow for air to be trapped in the cement paste during mixing. In comparison, the K-P-50 mix has also a large number of pores, but they are significantly smaller. The pulverisation of has resulted in refining the CBA particles to a comparable PC in size and SSA. As such, the combination of the pulverised CBA and the PC as a binder was much more homogenous in comparison to the K-O-50, resulting in less entrapped air in the mix. In addition, the formation of C-S-H gel in the K-P-50 sample has filled up the larger pores that have formed, resulting in finer pores. Similar observations were noted for TB-CBA (omitted in the paper).

5 Conclusions and Recommendations

In view of the presented results, it can be concluded that the CBA cement paste samples exhibited better strength improvements upon treatment of the raw CBA, particularly pulverisation. Pulverisation and secondly burning outperform soaking of the raw CBA for both sources examined in this study. This is caused by the increased SSA in the pulverised CBA and the removal of water-trapping carbon and other impurities in the burnt CBA. Optimisation of the CBA pulverisation can be performed by grinding CBA to higher fineness and larger SSA than in this study. The ideal SSA would be similar to the SSA of PC (approx. 350 m^2/kg).

The source of coal (composition) has a noticeable effect on CBA performance in a cementitious matrix. Owing to its higher silica content, the majority of the pulverised and burnt K-CBA samples displayed higher compressive strength than of TB-CBA. The original and soaked TB-CBA samples had higher SSA than the K-CBA samples, resulting in better compressive strength for those mixes. It is economically viable to increase CBA replacement level for industrial applications. The 50% CBA mixes have the lowest compressive strength of all the smaller CBA replacements. However, the pulverised and burnt CBA mixes from both sources showed acceptable strengths for basic concrete applications, such as pavers.

Consideration to combined treatments, e.g. pulverisation and burning, could possibly provide higher compressive strengths. However, from an economic perspective, pulverising and burning could be cost and energy intensive. Therefore, optimisation of CBA treatment process while maintaining an appealing pozzolanic activity must be considered. Utilisation of affordable additives such as superplasticiser or alkali activators are important and should be simultaneously investigated. This could be coupled with reducing l/b ratio to 0.25-0.3 to improve workability and attain higher compressive strengths.

Acknowledgements. The authors would like to thank Mr. Calvin Ooi from Sunway Paving Solutions and Mr. G. Sivakumar from the Sultan Salahuddin Abdul Aziz Power Plant in Kapar, Malaysia for their assistance in collecting CBA. We would like to extend our gratitude to Lahiru Gunawardena, Ze Ching Liew and Wong Yee Zhe for their assistance and other contributions to the experimental work. This research was made possible through the 2018 Monash University Malaysia – Sunway Group of Companies Grant Scheme.

References

American Coal Ash Association. CCP FAQs. American Coal Ash Association (2017). https://www.acaa-usa.org/aboutcoalash/ccpfaqs.aspx#Q13. Accessed 8 Jan 2019

Argiz, C., Sanjuán, M., Menéndez, E.: Coal bottom ash for Portland cement production. Adv. Mater. Sci. Eng. **2017**, 1–7 (2017). https://doi.org/10.1155/2017/6068286

ASTM C204-18e1. Standard Test Methods for Fineness of Hydraulic Cement by Air-Permeability Apparatus, American Society for Testing and Materials. West Conshohocken, PA (2018). https://doi.org/10.1520/c0204-07.2

ECOBA. Production and utilisation of CCPs in Europe (EU 15) in 2016, ECOBA European Coal Combustion Products Association (2016). http://www.ecoba.com/ecobaccpprod.html. Accessed 22 Jan 2019

Ibrahim, M.H.W., et al.: Split tensile strength on self-compacting concrete containing coal bottom ash. Procedia Soc. Behav. Sci. **195**, 2280–2289 (2015). https://doi.org/10.1016/j.sbspro.2015.06.317. Elsevier

Jang, J.G., et al.: Resistance of coal bottom ash mortar against the coupled deterioration of carbonation and chloride penetration. Mater. Design **93**, 160–167 (2016). https://doi.org/10.1016/j.matdes.2015.12.074. Elsevier

Latifi, N., et al.: Strength and physico-chemical characteristics of fly ash–bottom ash mixture. Arab. J. Sci. Eng. **40**(9), 2447–2455 (2015). https://doi.org/10.1007/s13369-015-1647-4

MATLAB R2017b. Natick, Massachusetts: The Mathworks, Inc. (2017)

Morse, E., Turgeon, A.: Coal, National Geographic Society (2012). https://www.nationalgeographic.org/encyclopedia/coal/. Accessed 12 Dec 2018

Naganathan, S., Mohamed, A.Y.O., Mustapha, K.N.: Performance of bricks made using fly ash and bottom ash. Constr. Build. Mater. **96**, 576–580 (2015). https://doi.org/10.1016/j.conbuildmat.2015.08.068. Elsevier

Oruji, S., et al.: Strength activity and microstructure of blended ultra-fine coal bottom ash-cement mortar. Constr. Build. Mater. **153**, 317–326 (2017). https://doi.org/10.1016/j.conbuildmat.2017.07.088

Rafieizonooz, M., et al.: Investigation of coal bottom ash and fly ash in concrete as replacement for sand and cement. Constr. Build. Mater. **116**, 15–24 (2016). https://doi.org/10.1016/j.conbuildmat.2016.04.080. Elsevier

Development of Porous Concrete with Recycled Aggregate

Saroj Mandal(✉) and Bibekananda Mandal

Jadavpur University, Kolkata, India
mailtosarojmandal@rediffmail.com

Abstract. An attempt has been made to develop porous concrete with recycled concrete aggregate for practical use. The mechanical properties of porous concrete with fully natural aggregate, porous concrete with fully recycled aggregate and porous concrete with both natural and recycled (50:50) aggregate have been studied. Recycled concrete aggregates were derived from the old crushed concrete cubes (already tested). The coefficient of water permeability by constant head method for concrete with natural aggregate, with recycling aggregate and with both aggregate (50:50) using in the range of 0.97 mm/s to 1.22 mm/s which is high enough to used as a drainage pavement for ground water recharge. The compressive, flexural and tensile strength of porous concrete with recycled concrete aggregate is comparatively less than that of porous concrete with natural aggregate of similar mixture. However, the deficiency can be improved by the use of a combination of natural and recycled aggregate (50:50).

1 Introduction

Porous concrete is one of the most effective pavement materials to address a number of important environmental issues, such as recharging groundwater and reducing storm water runoff. It has a large volume of air voids compared conventional concrete. For maximizing the benefit of its water permeability, several studies have been conducted to reveal the relationship between pore features and the hydraulic or acoustic conductivity of porous concrete using natural crushed aggregate (NA).

On the other hand, increase in the quantity of construction by-product, continuing shortage of dumping sites, sharp increase in transportation and disposal cost, stringent environmental pollution and regulation control guides to search for recycled aggregate instead of natural aggregate. The recycled Concrete aggregate (RCA) generally obtained from crushing concrete waste and subsequent processing (such as sieving, washing, separation and blending). The most notable feature of recycled aggregate is the component of attached cement mortar with the originally aggregate. The percentage of attached mortar depends on the size of the original aggregate. Smaller the size more will be the attached mortar. The findings of an experimental investigation on properties of pervious concrete with NA, RCA and a combination of NA & RCA are reported and discussed. The properties of various porous concrete samples including density, porosity, sorptivity, compressive strength, water permeability have been carefully measured. The 28-day compressive strength for pervious concrete with NA was reduced by the use of recycled concrete aggregate fully. However a combination of

M. Shehata et al. (Eds.): GeoMEast 2019, SUCI, pp. 22–29, 2020.
https://doi.org/10.1007/978-3-030-34249-4_3

recycled aggregate and natural aggregate (50:50) is more suitable in terms of strength and permeability for practical purposes.

2 Experimental Program

2.1 Material

Portland Slag Cement as per IS 455:2015 is used for different mixtures of porous concrete. Two types of coarse aggregates (a) Natural coarse aggregate (NA) (b) Recycled concrete aggregate (RCA) are used for making the porous concrete. NA consists of crushed coarse aggregate of 12.5 mm passing and retained on 10 mm sieve having specific gravity 2.60 and water absorption of 0.5% has been used. RCA has been derived from the crushed concrete cube (already tested) by hammering and subsequent sieving. The size fraction of RCA and NA has been kept similar. The specific gravity and water absorption of RCA are 2.50 and 4.80% respectively. The amount of mortar attached to the aggregate in RCA is 3.5%.

2.2 Mix Proportions

Total nine numbers of porous concrete mixtures have been made depending on the type of coarse aggregate, water cement ratio (W/C) and cement aggregate ratio (C/A) as shown in Table 1. The mixture no M1N to M3N represents the porous concrete mix with NA, M1R to M3R represents the porous concrete mix with RCA and M1NR to M3NR represents the porous concrete mix with 50% NA and 50% RCA. The cement aggregate ratio is kept as 1:4 for all mixtures. For each mixture, cube compressive strength (100 × 100 × 100 mm) at the age of 3 days, 7 days, 28 days, split tensile strength (200 mm height and 100 mm diameter) at 28 days and flexural strength (100 mm × 100 mm × 500 mm) at 28 days were measured. The water permeability test, sorptivity test were also made on porous concrete cubes at the age of 28 days. All the specimens are demoulded after 24 h casting and immersed in water to continue their curing until reaching testing age. It may be mentioned here that it is difficult to obtain the slump value in porous concrete with NA, RCA or 50% NA and 50% RCA. A collapsed slump has been obtained in all mixtures due to low binding property or weak bond in between cement and aggregate. However, this generally does not make any difficulties in compaction. All the test results are based on the six specimens for a particular mixture.

Table 1. Mixture proportion and their properties

Sl no.	Mixture no.	W/C ratio	Density (kg/m^3)	Porosity (%)	Sorptivity (mm/min$^{0.5}$)	Water permeability (mm/s)	Remarks
1	M1N	0.35	2020	18	2.90 E−04	1.06	NA (100%)
2	M2N	0.33	2050	16	2.32 E−04	1.04	
3	M3N	0.30	2110	14	2.21 E−04	0.97	
4	M1R	0.35	1930	21	3.61 E−04	1.16	RCA (100%)
5	M2R	0.33	1950	20	3.39 E−04	1.10	
6	M3R	0.30	1970	19	3.25 E−04	1.02	
7	M1NR	0.35	1940	19	2.97 E−04	1.04	NA and RCA (50:50)
8	M2NR	0.33	1965	17	2.81 E−04	1.02	
9	M3NR	0.30	2000	16	2.58 E−04	1.00	

3 Results and Discussion

Figure 1 shows the variation of density of different porous concrete mixtures. As usual the density of porous concrete mixes decreases with increases in water cement ratio. Further with the replacement of NA with RCA, the density is reduced for other parameters remain same as the density of RCA is less compared to NA. Figure 2 exhibits the relationship between porosity of different porous concrete vs. water cement ratio at cement aggregate ratio of 1:4. The porosity generally increases with increase of water cement ratio for all the three types of mixtures of porous concrete. This is due to presence of more void space in cement paste with higher water cement ratio. It is also noted that porous concrete with RCA has higher porosity than that of porous concrete with NA and porous concrete with 50% NA & 50% RCA. This is due to the presence of old weak attached mortar and more water absorption capacity in recycled concrete aggregate. The sorptivity of different mixtures of porous concrete are also shown Fig. 3. A linear relationship between sorptivity values with water cement ratio for porous concrete with both NA and RCA.

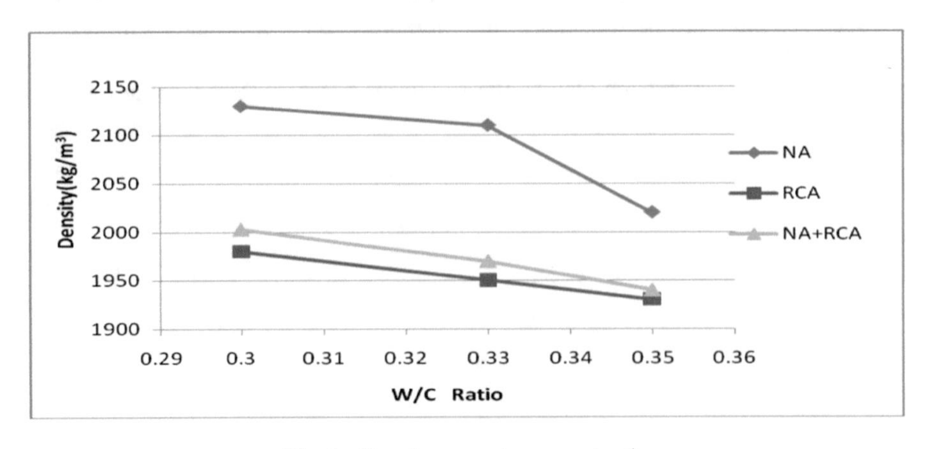

Fig. 1. Density vs. water cement ratio

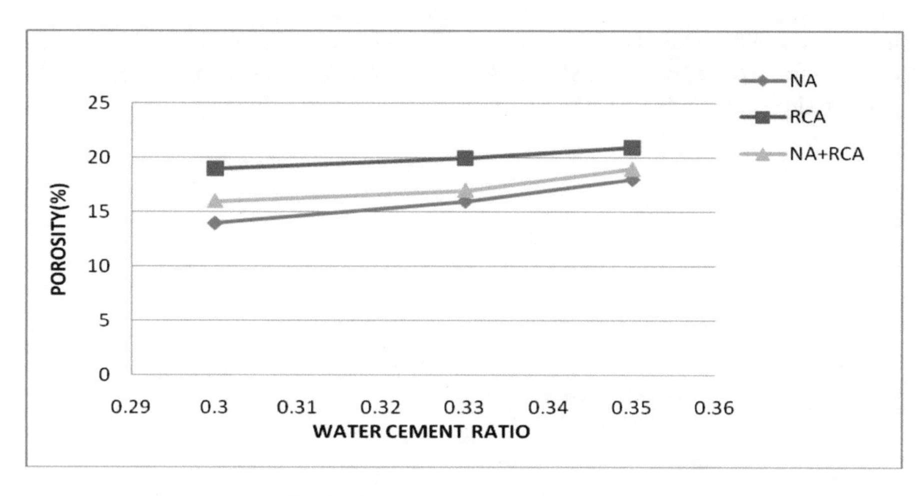

Fig. 2. Porosity vs. water cement ratio

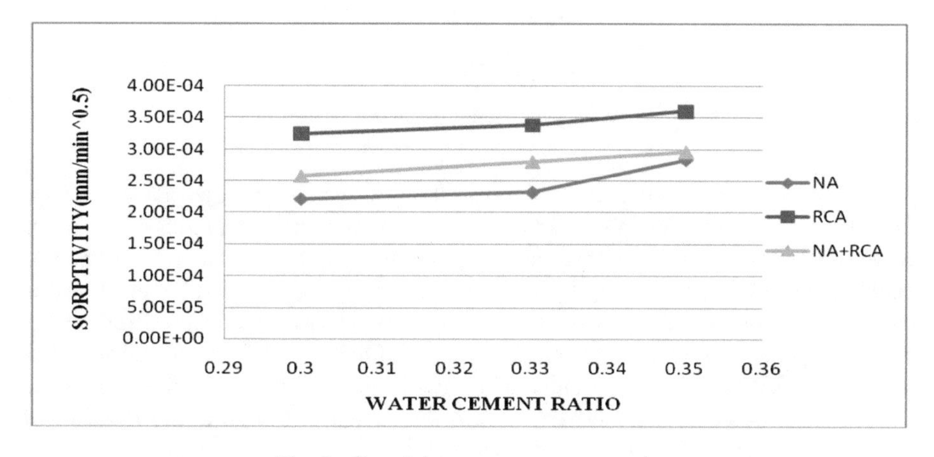

Fig. 3. Sorptivity vs. water cement ratio

The coefficient of water permeability determined by constant head method for each of porous concrete mixtures are shown Fig. 4. The maximum permeability achieved in porous concrete with NA, RCA and with 50% NA & 50% RCA are 1.14 mm/s, 1.16 mm/s and 1.22 mm/s respectively. For all the mixtures of different aggregate of porous concrete, the coefficient of water permeability is always greater than 1 mm/s, which is highly porous material for drainage purposes to recharge the ground water table. A linear relationship observed between the water permeability with water cement ratio for cement aggregate ratio 1:4 (Ref Fig. 4). The chart exhibits the coefficient of water permeability of porous concrete mixtures increases with increase in water cement ratio. This is due to more void space with increase in water cement ratio. It is also noted that the permeability of porous concrete with RCA is always greater than that of porous concrete with NA and porous concrete with 50% NA and 50% RCA. This is due to the weak attached mortar paste formation and more water absorption capacity of recycled concrete aggregate.

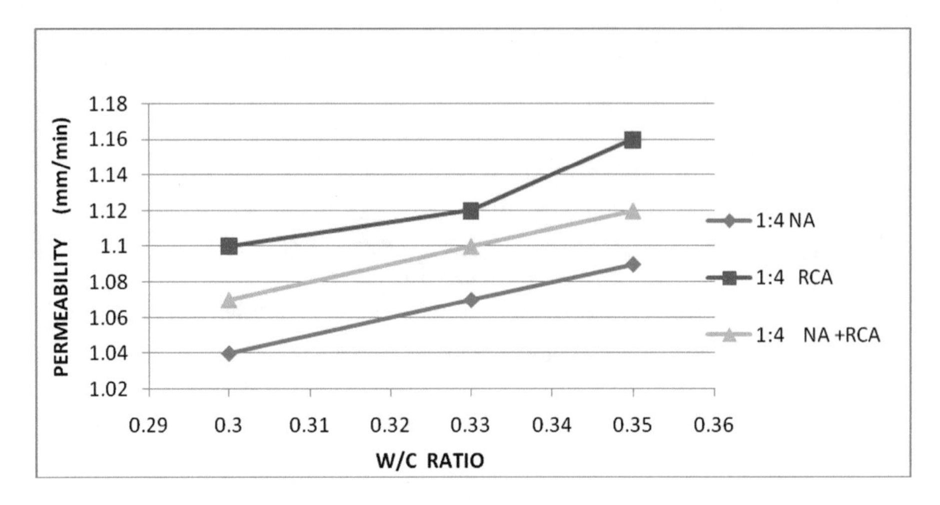

Fig. 4. Water permeability vs. water cement ratio

The results of cube compressive strength (at 3, 7, and 28 days), tensile strength (at 28 days) and flexural strength (at 28 days) of all the mixtures for porous concrete are presented in Table 2. As expected, the compressive strength of all the mixes of porous concrete increases with ages up to 28 days. The cube compressive strength of porous concrete with NA is increases with the decrease in water cement ratio also (Ref Fig. 5). Similar trend is also noticed for porous concrete with RCA (Ref Fig. 6) and with NA and RCA (50:50) (Ref Fig. 7). With the increase in water cement ratio, the voids in the cement paste becomes more thereby reduce the strength of paste at the interface of aggregate and paste. However for a particular W/C ratio, the cube compressive strength of porous concrete with RCA is less than porous concrete with NA at all ages. This reduction is mainly due to the presence of weak mortar attached to the aggregate. Again the mixtures of NA and RCA (50:50) for porous concrete show the compressive strength in between the compressive strength of porous concrete with fully NA and fully RCA. It is noted that the result of tensile and flexural strength of porous concrete mixtures follow the similar trends as in the case of compressive strength of porous concrete cube at 28 days.

Table 2. Compressive Strength, Tensile Strength and Flexural Strength of various mixtures of porous concrete

Mix no.	Compressive strength (MPa)			Tensile strength (MPa)	Flexural strength (MPa)
	3 days	7 days	28 days		
M1N	6.52 ± 0.12	7.35 ± 0.20	10.88 ± 0.60	0.98 ± 0.16	2.00 ± 0.42
M2N	7.50 ± 0.22	9.77 ± 0.10	13.80 ± 0.52	1.10 ± 0.19	2.40 ± 0.34
M3N	8.20 ± 0.18	11.20 ± 0.54	14.70 ± 0.63	1.16 ± 0.32	2.60 ± 0.40
M1R	5.12 ± 0.32	5.45 ± 0.32	6.75 ± 0.56	0.70 ± 0.18	1.20 ± 0.20
M2R	5.45 ± 0.14	6.12 ± 0.24	7.45 ± 0.45	0.88 ± 0.22	1.50 ± 0.10
M3R	6.00 ± 0.25	7.50 ± 0.58	9.40 ± 0.67	1.02 ± 0.30	1.80 ± 0.8
M1NR	6.00 ± 0.02	7.10 ± 0.32	9.80 ± 0.50	0.80 ± 0.16	1.80 ± 0.22
M2NR	6.50 ± 0.19	8.00 ± 0.19	12.20 ± 0.42	0.95 ± 0.25	2.20 ± 0.24
M3NR	7.10 ± 0.30	9.40 ± 0.42	13.70 ± 0.70	1.05 ± 0.34	2.40 ± 0.22

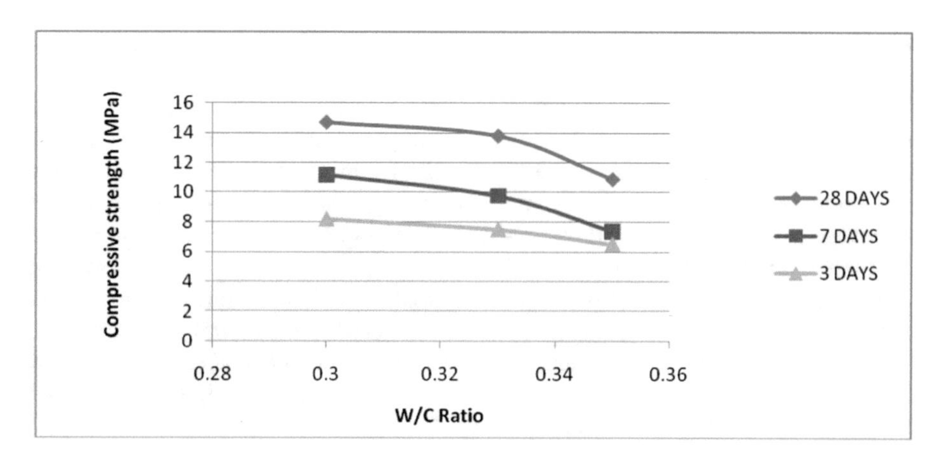

Fig. 5. Compressive strength vs. water cement ratio at different ages

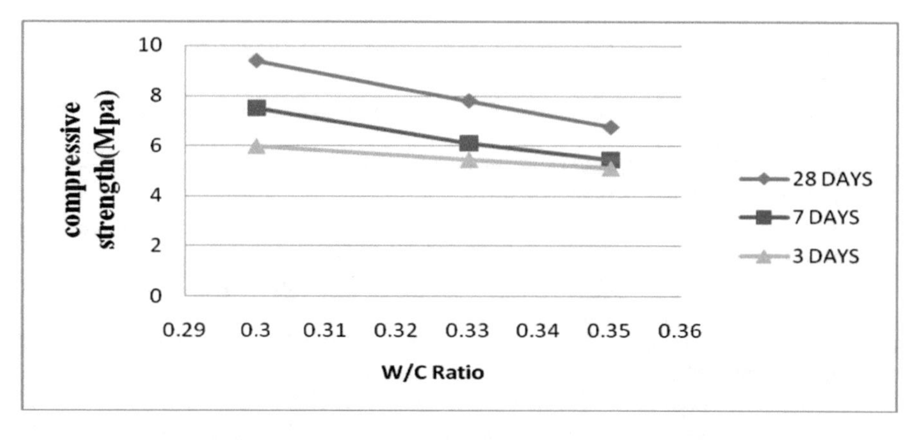

Fig. 6. Compressive strength vs. water cement ratio at different ages

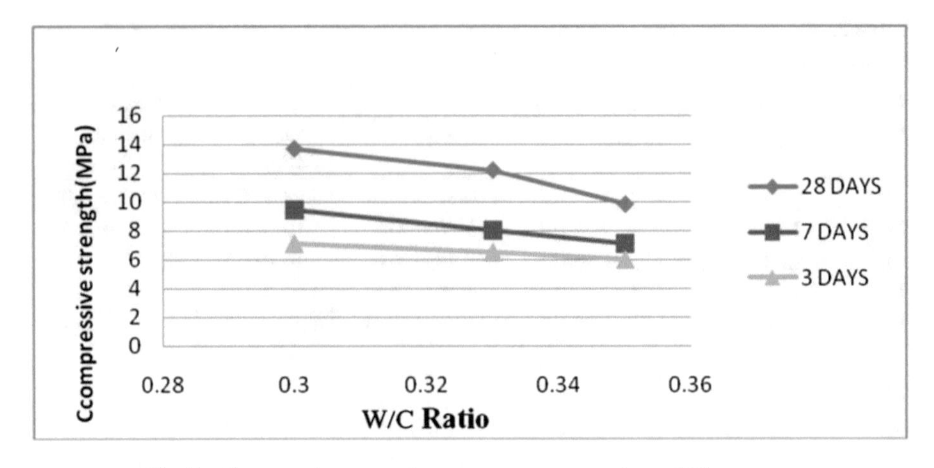

Fig. 7. Compressive strength vs. water cement ratio at different age.

4 Conclusion

Porous concrete with RCA shows low density and higher porosity than porous concrete with NA. The coefficient of water permeability of porous concrete with NA, RCA and both NA and RCA (50:50) are in the range of 0.97 mm/s to 1.16 mm/s, 1.02 mm/s to 1.22 mm/s and 1.00 mm/s to 1.10 mm/s respectively, which is high enough to used as a drainage pavement for ground water recharge. In Sorptivity test, the maximum value is achieved in porous concrete with RCA. This is due to more water absorption capacity in RCA than that of NA. The compressive strength of different mixtures of porous concrete decrease with the increase of water cement ratio. Porous concrete with RCA show lower compressive strength, tensile strength and flexural strength than that of porous concrete with NA. However, the use of 50% NA and 50% RCA in porous concrete is more suitable considering all aspects.

References

ACI Committee 211: Guide for Selecting Proportions for No-Slump Concrete, pp. 1–26. American Concrete Institute, Farminton Hills (2002)

ACI Committee 522: Report on Porous Concrete, pp. 1–38. American Concrete Institute, Farmington Hills (2002)

Andrew, I.N., Bradley, J.P.: Effect of aggregate size and gradation on porous concrete mixtures. ACI Mater. J. **2010**, 625–631 (2010)

Bury, M., Mawby, C.: Porous concrete mixes: the right ingredients and proportions are critical to success. Concr. Constr. World Concr. **51**(4), 37–40 (2006)

Ghafoori, N., Dutta, S.: Building and non pavement applications of no-fines concrete. J. Mater. Civil Eng. **7**(4), 286–289 (1995)

Mandal, S., et al.: Some studies on durability of recycled aggregate concrete. Indian Concr. J. **76**(6), 385–388 (2002)

Mandal, S., et al.: Strength and durability of recycled aggregate concrete. In: IABSE Symposium Report, vol. 86, no. 6, pp. 33–46 (2002)

Ghafoori, N.: Development of no-fines concrete pavement applications. J. Transp. Eng. **126**(3), 283–288 (1995)

Yang, J., Jiang, G.: Experimental study on properties of porous concrete pavement materials. Cem. Concr. Res. **33**, 381–386 (2003)

Kevern, J.T., et al.: Porous concrete mixture proportions for improved freeze-thaw durability. J. ASTM Int. **5**(2), 1–12 (2008)

Katz, A.: Properties of concrete made with recycled aggregate from partially hydrated old concrete. Cem. Concr. Res. **33**, 703–711 (2003)

Shao, Y., Lin, X.: Early-age carbonation curing of concrete using recovered CO_2. Concr. Int. **33**, 50–56 (2011)

Soil Conservation on Slopes Subject to Water Erosion: The Application of the Concrete Lozenges Channels Technique for Slope Stability

Latifa El Bouanani(✉) and Khadija Baba(✉)

GCE Laboratory, High School of Technology-Sale, Mohammed V University, Rabat, Morocco
elbouanani.latifa@gmail.com, khadija_baba@hotmail.com

Abstract. The concrete lozenges are a new technique used to protect slopes. It constitutes a non-continuous mesh mask on the slop, having the effect of collecting and transporting runoff water. Therefore, it allows on the one hand, reducing the quantities of soil loss in the shallow surface, and on the other hand, to minimize the water stagnation time and their infiltration at a depth which will influence the soil shear strength causing significant sliding mode of slope failure. The number and geometry of the lozenges depend of both natural and anthropogenic parameters such as: the nature of the soil, the rainfall zone, the quantity of soil that can be removed by the channel located at the bottom of the slope and the slope geometry. A parametric study is undertaken to define the optimal design of concrete lozenges.

Keywords: Water erosion · RUSLE · Concrete lozenges · Area at risk · Slope length

1 Introduction

Water erosion is a natural phenomenon, which is defined as the total process of detachment, transport and deposition of solid particles from the soil surface due to rainfall, runoff or both. The risk of water erosion occurs in both drainage basins and road and rail slopes. It can also occur when the runoff capacity is greater than the infiltration capacity due to soil saturation or to the formation of a bedding layer, named soil crusting, which minimizes the infiltration capacity of the soil. Also, according to Horton's infiltration model, if the rainfall supply exceeds the infiltration capacity, infiltration tends to decrease in an exponential manner. Indeed, this phenomenon can be aggravated causing major gullies and landslides [1].

In accordance with Ennadifi the Rif in general and the Prerif in particular are zones characterized by the predominance of friable lithological formations such as marls, marl-limestones and schists [2, 3]. The rains in the Rif and the Prerif are characterized by limited rains in time and space, but which give rise to very large amounts of water in a few days, or even in just a few hours with very high instantaneous intensities. They have a direct effect on the erosion mechanisms and floods [3, 4]. In addition, slopes

M. Shehata et al. (Eds.): GeoMEast 2019, SUCI, pp. 30–42, 2020.
https://doi.org/10.1007/978-3-030-34249-4_4

surrounding the Taourirte-Nador railway line, more precisely PK 94, due to their location in the oriental Rif, are subject to intense erosion, which necessitates periodic maintenance, generating significant costs.

According to Roose [5], as long as the infiltration has not improved on the drainage basin, we should not attempt to refill the gully (otherwise it will find another bed), but provide a stable channel able to evacuate flows of the decadal flood (minimum). In doing so, concrete lozenges are a preventive technique which proposes the creation of well-developed artificial gullies, notably concrete lozenges to frame the surface of exposed soil and to direct the runoff paths towards these gullies (Fig. 1).

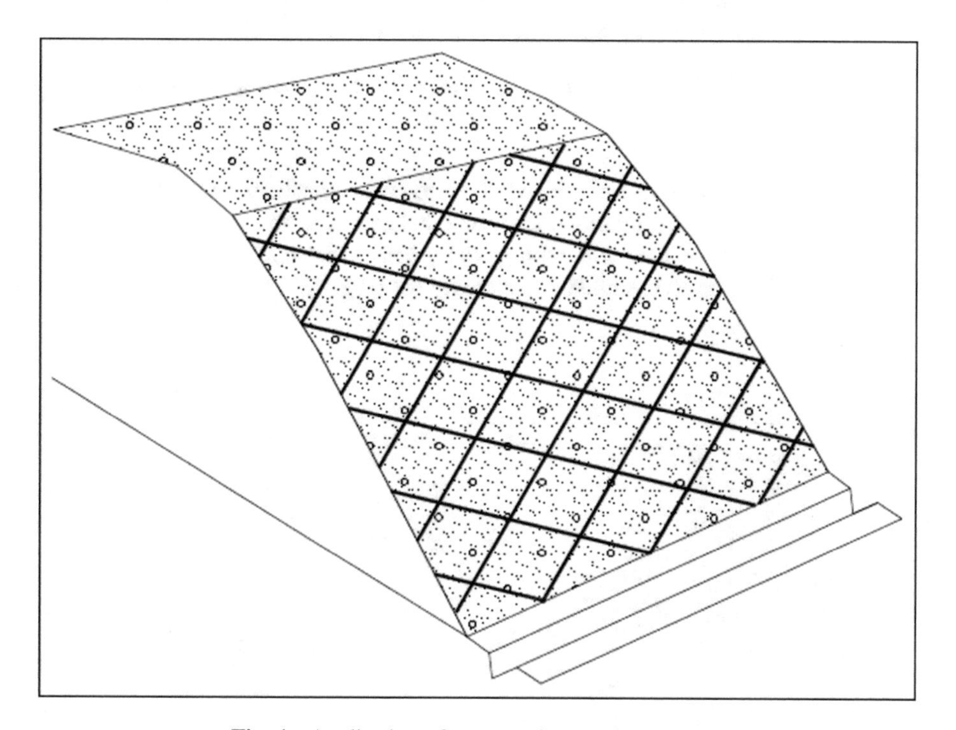

Fig. 1. Application of concrete lozenges' technique

In discussing concrete lozenges impact on the slope erodibility, we will apply the Revised Universal Soil Loss Equation RUSLE to quantify the quantities of soil eroded before and after the use of this method.

2 Geological and Climatological Framework

In discussing the concrete lozenges geometry, the PK 94's slope of the Taourirte-Nador railway line located in the North-West of Morocco in the Nador city in the oriental Rif, which is subject to intense water erosion (Figs. 2 and 3) is taken as a study case.

Fig. 2. Gullying in the slope crest

Fig. 3. Filling of drainage channels by sediment

2.1 Geology

In the geological context, the oriental Rif presents formations characterized by a superior Miocene, and an inferior Pleistocene belonging to the villafranchian. These lands consist mainly of pink silts with gravelly beds and a small band of a powerful quaternary; belonging to the middle quaternary [6] (Fig. 4).

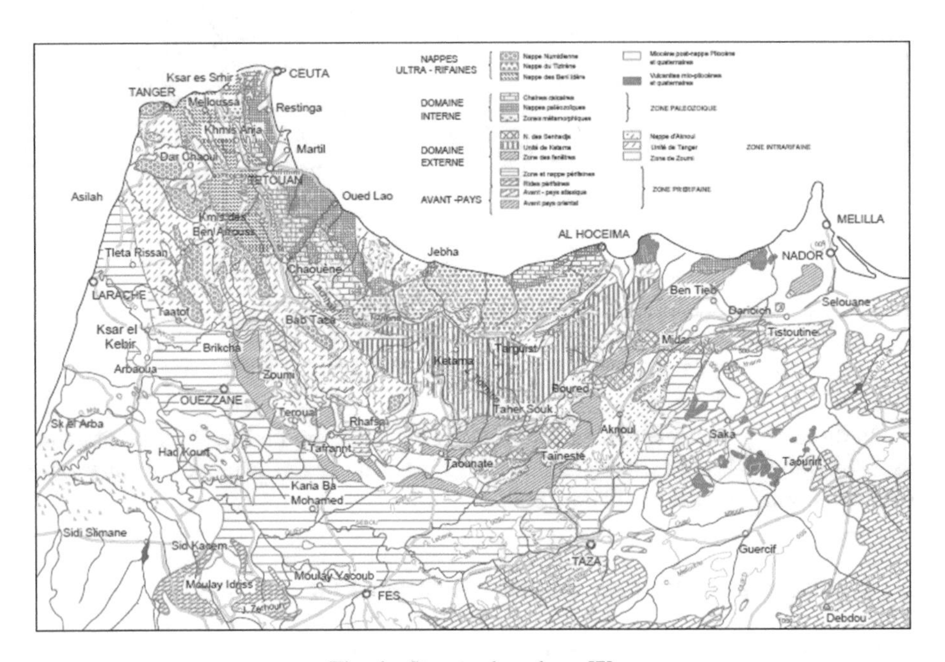

Fig. 4. Structural geology [7]

2.2 Granulometric Characteristics

The soil is characterized by a strong dominance of silt-clay formations and fine sand [6]:

- The silt-clay formations have an average of 90% of particles with a diameter of less than 80 μm and 10% of fine sand.
- The fine sand formation comprises an average of 25% of particles with a diameter of less than 80 μm.

2.3 Precipitations

The province of Nador is characterized by a Mediterranean climate, including alternating dry and wet seasons, from June to September, and from October to April. The average annual rainfall is approximately 373,5 mm/year. Rains are characterized by seasonal variability.

The following Fig. 5 gives the average of total monthly precipitation in mm of 12 years (2001 to 2012) [6]:

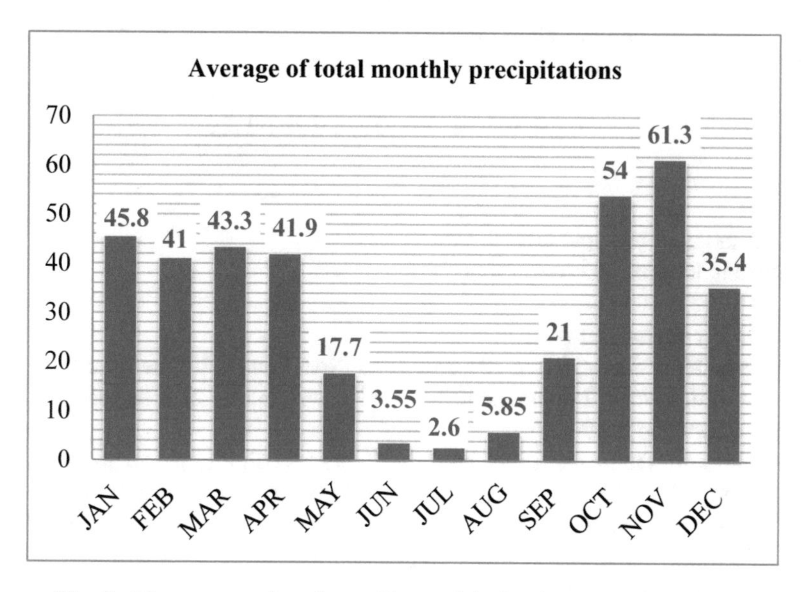

Fig. 5. The average of total monthly precipitation in mm (2001 to 2012)

3 Methods

3.1 Methods

RUSLE is an erosion model designed to predict the long-time average annual soil loss (A) carried by runoff from specific field slopes in specified cropping and management systems as well as from rangeland. It is also applicable to non-agricultural conditions such as construction sites [8].

Soil erosion can be calculated using the Revised Universal Soil Loss Equation such as:

$$A = R * K * LS * C * P \quad (1)$$

With:

- A = Predicted soil loss (ton.ha-1.year-1)
- R = Rainfall and runoff factor (MJ.mm/ha.h.year);
- K = Soil erodibility factor (t.h/MJ.mm);
- LS = Slope length and steepness factor (a dimensionless factor);
- C = Crop management factor (a dimensionless factor);
- P = Support practices factor (a dimensionless factor).

Rainfall and runoff factor (R)

The rainfall and runoff factor is the average annual of the sum of the storm EI value; which is the product of total storm energy by the maximum 30-min intensity (I30) (MJ. Mm/ ha.h. y).

Given that, we have only the annual and monthly precipitation amounts data, we will use the Rango and Arnoldus equation [9], for the definition of the R factor.

Therefore, the R factor can be expressed as follows:

$$\text{Log } R = 1{,}74 * \text{Log } \Sigma\left(Pi^2/P\right) + 1{,}29 \quad (2)$$

With:

- Pi: the monthly precipitation in mm;
- P: the annual precipitation in mm.

Soil erodibility factor (K)

The soil-erodibility factor (K) is the rate of soil loss caused by the effect of rainfall, runoff and infiltration. It's measured on a plot which has 22.1 m long, a 1.83 m minimum width and a 9% slope.

Soil-erodibility factor can be defined by the soil-erodibility nomograph, or by the algebraic approximation used when the silt fraction does not exceed 70% [8].

Both the graphic and the algebraic method are based on five parameters: % modified silt, % modified sand, % organic matter (OM), classes for structure (s) and permeability (p).

K can be calculated using the algebraic approximation such us [8, 10]:

$$K = \frac{\left[2.1 * 10^{-4}(12 - OM)M^{1.14} + 3.25(s - 2) + 2.5(p - 3)\right]}{100} \quad (3)$$

M is the product of the primary particles size:

$$M = (\%\text{modified silt or the } 0.002 - 0.1\,\text{mm size fraction}) \times (\%\text{silt} + \%\text{sand}) \quad (4)$$

To defining the s factor, a nomographic classification of the soil structure based on the particles size [8], it is given in the following Table 1:

Table 1. Soil structure's classification

Class	Type of soil structure	Size (mm)
1	Very fine granular or without structure	–
2	Fine granular	<2
3	Medium granular Coarse granular	2–5 5–10
4	Polyhedral, lamellar, massive, prismatic	>10

According the soil structure and the soil permeability, Cook et al. [11] are defined six permeability class. They are presented in the following (Table 2):

Table 2. Permeability class [11]

Structural class	Permeability class	Permeability cm/s	Permeability code
Gravel, coarse sand	Fast	>4,4.10–3	1
Loam sand and loams sandy	Moderate to fast	(1,4–4,4).10–3	2
Fine sand loams, loams	in moderation	(0,4–1,4).10–3	3
Loams, clay loams	Slow to moderate	(0,14–0,4).10–3	4
Clay loams, clays	slow	(4–14).10–5	5
Tight, compacted	Very slow	<4.10–5	6

Slope length and steepness factor (LS)
LS is a dimensionless factor that represents the topography effect on erosion. LS combines both slope length factor (L in m) and the slope steepness factor (S in %).

1. Slope length factor (L)
   Slope length factor can be calculated using the following equation:

$$L = (l/72.6)^{m} \tag{5}$$

Where l is the horizontal projection, 72.6 (ft) is the RUSLE unit plot length and m is a dimensionless variable slope length exponent [8, 10].

$$m = \beta/(1 + \beta) \tag{6}$$

To calculate the slope length exponent (m) we should defining the β variable [8, 12] by the following equation:

$$\beta = (\frac{\sin\theta}{0.0896})/(3 * (\sin\theta)^{0.8} + 0.56) \tag{7}$$

This value for the ration β of rill to interril erosion for conditions when the soil is moderately susceptible to both rill and interill erosion were computed from the Eq. 7 [8].Where θ is the slope angle.

2. Slope steepness factor (S)
   The slope steepness factor S is defined by the following equation [13]:

$$S = 10.8 \sin\theta + 0.03 \quad s < 9\% \tag{8}$$

$$S = 16.8 \sin\theta - 0.5 \quad s \geq 9\% \tag{9}$$

LS can be defined also using the following table using the slopes angles in % and the slopes length in ft:

Table 3. Slope length and steepness factor (LS) [8]

Slope %	Slop length in feet																
	<3	6	9	12	15	25	50	75	100	150	200	250	300	400	600	800	1000
0,20	0,05	0,05	0,05	0,05	0,05	0,05	0,05	0,05	0,05	0,05	0,06	0,06	0,06	0,06	0,06	0,06	0,06
0,50	0,07	0,07	0,07	0,07	0,07	0,07	0,08	0,08	0,09	0,09	0,10	0,10	0,10	0,11	0,12	0,12	0,13
1,00	0,09	0,09	0,09	0,09	0,09	0,10	0,13	0,14	0,15	0,17	0,18	0,19	0,20	0,22	0,24	0,26	0,03
2,00	0,13	0,13	0,13	0,13	0,13	0,16	0,21	0,25	0,28	0,33	0,37	0,40	0,43	0,48	0,56	0,63	0,69
3,00	0,17	0,17	0,17	0,17	0,17	0,21	0,30	0,36	0,41	0,50	0,57	0,64	0,69	0,80	0,96	1,10	1,23
4,00	0,20	0,20	0,20	0,20	0,20	0,26	0,38	0,47	0,55	0,68	0,79	0,89	0,98	1,14	1,42	1,65	1,86
5,00	0,23	0,23	0,23	0,23	0,23	0,31	0,46	0,58	0,68	0,86	1,02	1,16	1,28	1,51	1,91	2,25	2,55
6,00	0,26	0,26	0,26	0,26	0,26	0,36	0,54	0,69	0,82	1,05	1,25	1,43	1,60	1,90	2,43	2,89	3,30
8,00	0,32	0,32	0,32	0,32	0,32	0,45	0,70	0,91	1,10	1,43	1,72	1,99	2,24	2,70	3,52	4,24	4,91
10,00	0,35	0,37	0,38	0,39	0,40	0,57	0,91	1,20	1,46	1,92	2,34	2,72	3,09	3,75	4,95	6,03	7,02
12,00	0,36	0,41	0,45	0,47	0,49	0,71	1,15	1,54	1,88	2,51	3,07	3,60	4,09	5,01	6,67	8,17	9,57
14,00	0,38	0,45	0,51	0,55	0,58	0,85	1.40	1,87	2,31	3,09	3,81	4,48	5,11	6,30	8,45	10,40	12,23
16,00	0,39	0,49	0,56	0,62	0,67	0,98	1,64	2,21	2,73	3,68	4,56	5,37	6,15	7,60	10,26	12,69	14,96
20,00	0,41	0,56	0,67	0,76	0,84	1,24	2,10	2,86	3,57	4,85	6,04	7,16	8,23	10,24	13,94	17,35	20,57
25,00	0,45	0,64	0,80	0,93	1,04	1,56	2,67	3,67	4,59	6,30	7,88	9,38	10,81	13,53	18,57	23,24	27,66
30,00	0,48	0,72	0,91	1,08	1,24	1,86	3,22	4,44	5,58	7,70	9,67	11,55	13,35	16,77	23,14	29,07	64,71
40,00	0,53	0,85	1,13	1,37	1,59	2,41	4,24	5,89	7,44	10,35	13,07	15,67	18,17	22,95	31,89	40,29	48,29
50,00	0,58	0,91	1,31	1,62	1,91	2,91	5,16	7,20	9,13	12,75	16,16	19,42	22,57	28,60	39,95	50,63	60,84
60,00	0,63	1,07	1,47	1,84	2,19	3,36	5,97	8,37	10,63	14,89	18,92	22,78	26,51	33,67	47,18	59,93	72,15

Crop management factor (C)

Particular attention is also paid to cultural practices such as the crops orientation. Cultures parallel to the slope offer little retention and storage while the runoff is intense [14].

C is a dimensionless factor that represents the effect of cropping and management practices. The determination of the C factor is related to the density of the vegetation cover of the soil surface.

Support practices factor (P)

P factor, a dimensionless factor, is a ratio that expresses the effect of a specific support practice on the erosion. The P factor varies according to the slope (table) as well as the agricultural or anti-erosive practices adopted. The values of the factor P are less than or equal to 1. The value 1 corresponds to the ground without anti-erosion practices (Table 4).

Table 4. Value of support practices factor [15].

Slope steepness (%)	P
0.0–7.0	0.55
7.0–11.3	0.60
11.3–17.6	0.80
17.6–26.8	0.90
>26.8	1.00

3.2 Application of the Technique

Concrete lozenges are a technique which allows the conservation of the shallow soil and the slopes stability. They are inclined concrete drainage channels in the form of lozenges limiting the formation of deep runoff paths, as well **as the fact that** they collect and transport water down the slope to reduce, on the one hand, the flow energy of runoff which cause the transport of loosened soil and on the other hand, reduce the quantities of water infiltrated.

Using the Wishmeier model to quantify soil losses on the slopes, especially the Revised Universal Soil Loss Equation (RUSLE), we will discuss the parameters that influence the choice of lozenges geometry:

- The area occupied by concrete channels (the number of lozenges);
- The length of lozenges' diagonals (the slope length).

It should be noted that: the execution techniques and the constructive arrangements details will be the object of another paper with details of the lozenges stability.

The application of this new technique is studied on six cases to look for optimal number and geometry of these lozenges-meshes: a PK94's slope stability of the Taourirte-Nador railway line is studied by application of this technique to simulate the degree of slope protection. The six studied cases are:

Case n° 1 (area of 100 × 50 m^2 without protection)
It is a sloping surface, devoid of any means of protection. This study case is taken as a reference to compare the impact of the present technique on the slope stability. It has an area of 5000 m^2 under normal conditions of exposure to the impact of rainfall and runoff. This surface has for geometry 100 × 50 m^2.

Case n° 2 (Area of 100 × 50 m^2 with protection by a single lozenge)
On the same reference area of 5000 m^2 (100 × 50), an application of the new technique is made by studying the case of introduction of a single lozenge that divides the slope into a lozenge-mesh of respective diagonal's length of d1 = 50 m and d2 = 100 m. with d1 is the diagonal following the slope.

Case n° 3 (area of 100 × 50 m^2 with protection by two lozenges)
To simulate the impact of increasing the number of lozenges on the slope stability against water erosion, a study of the impact of the use of two equal diamonds whose diagonal lengths are: d1 = 25 m and d2 = 100 m.

In order to show the importance of the geometry of lozenges on the optimization of soil loss quantities, we study the variant of the introduction of concrete lozenges on an area of 5000 m^2 (50 × 100) initially unprotected.

Case n° 4 (Area of 50 × 100 m^2 without protection)
For this case we study the slope stability for the same sloping surface of the reference slope (5000 m^2), but of different geometry (50 × 100).

Case n° 5 (Area of 50 × 100 m^2 with protection by a single lozenge)
A study of the introduction of a single lozenge whose diagonal's length are d1 = 100 m and d2 = 50 m.

Case n° 6 (Surface of 50 × 100 m^2 with protection by two lozenges)
For the sixth case a study of the introduction of two lozenges in the respective diagonals' lengths are d1 = 100 m and d2 = 25 m.

The figure below schematizes the six studied cases (Fig. 6):

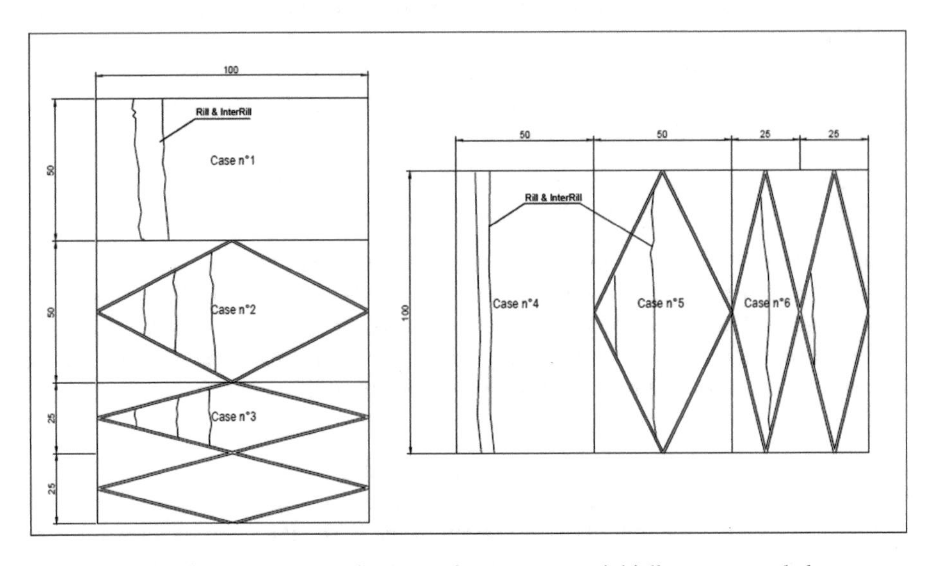

Fig. 6. The arrangement of concrete lozenges on an initially unprotected slop

4 Results and Discussions

4.1 Results

Rainfall and runoff factor (R)
Using the Eq. (2) R factor has the following value: R = 150 MJ.mm/ha.h.year

Soil erodibility factor (K)
K factor is computed according to the Eqs. (3 and 4). It has the following value: K = 0.107 mm.t.ha/MJ. (Table 5).

Table 5. K factor calculation

% clay	59	%
% Limon	30	%
% fin sand	10	%
M	1640	–
OM	1	%
s	2	–
p	3	–
K	**0.107**	**t.h/MJ.mm**

Slope length and steepness factor (LS)

We are discussing the change in the slope length factor per the change in the lozenges diagonal length for the six studied cases. The following graphic (Fig. 7) shows the results of the LS factor calculation according to the Table 3:

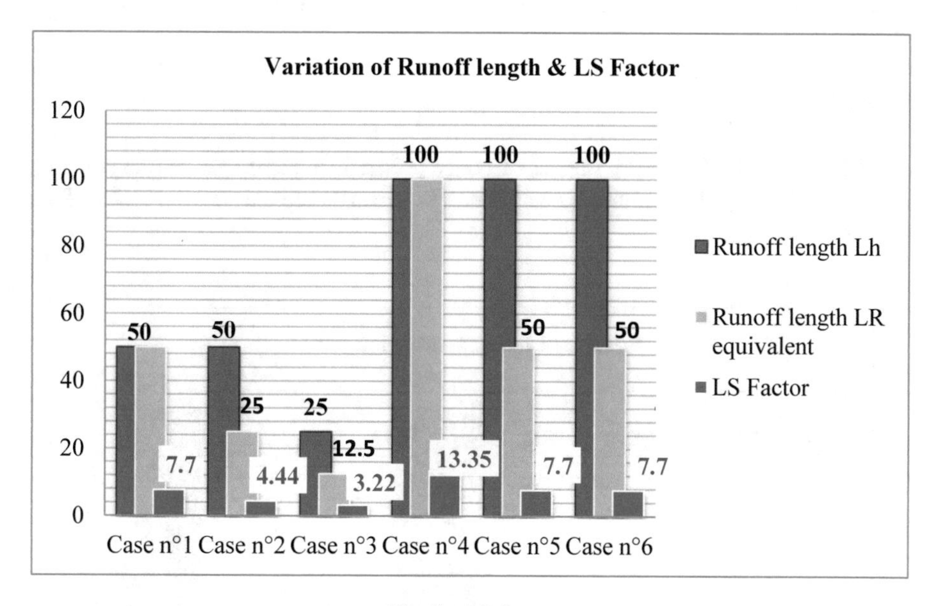

Fig. 7. LS factor

- Runoff length lh is the longest runoff distance in order to reach outlet by a quantity of water (the inclined channel for the case of lozenges-mesh).
- Runoff length equivalent Lr: Runoff length in the concrete lozenges-mesh is simulated at half of the length of lozenge's diagonal.

Predicted soil loss A

The Table 6 contains the data required for the Predicted soil loss calculation both before and after the use of lozenges technique for the six studied cases:

Table 6. Predicted soil loss (A)

Lozenges category	Case n°1	Case n°2	Case n°3	Case n°4	Case n°5	Case n°6	Unit
Exposed area	0,50	0,48	0,47	0,50	0,48	0,47	ha
LS factor	7,7	4,44	3,22	13,35	7,7	7,7	–
Erodibility factor (K)	0,107	0,107	0,107	0,107	0,107	0,107	mm.t.h/MJ
Erosivity factor (R)	150	150	150	150	150	150	MJ.mm/ha. h.year
The slope is devoid of vegetation cover and support practice (C = 1 and P = 1)	1	1	1	1	1	1	–
Predicted soil loss (A)	61,79	34,52	24,37	107,13	59,87	58,28	(ton.year-1)

5 Discussions

RUSLE is used to define the impact of concrete-lozenges technique on the soil loss. The following Fig. 8 summarizes variations between predicted soil loss (A) of the studies cases:

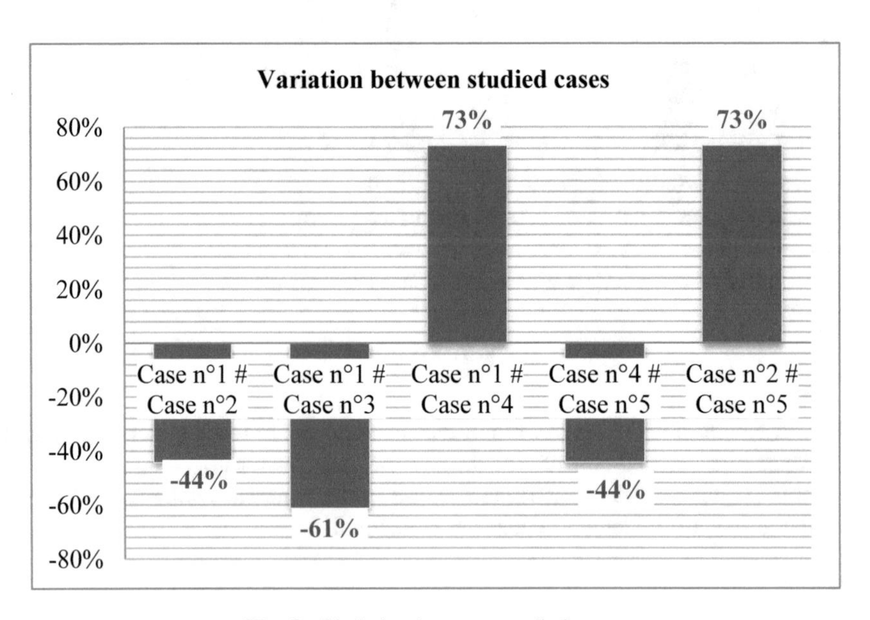

Fig. 8. Variation between studied cases

In discussing the variation between studied cases, we conclude that:

- Concrete lozenges technique reduces the soil loss by 44% for a single lozenge (Case n° 1 # Case n° 2 & Case n° 4 # Case n° 5);
- As the lozenges number goes up, the soil loss become less by 61% for two lozenges (Case n° 1 # Case n° 3);
- The reduction in sol loss is linked and conditional on the lozenges geometry: for the same area surrounded by concrete channels we can reduce erosion by 73% if we choose the adequate geometry. (Case n° 2 # Case n° 5).

6 Conclusion

According to RUSLE, this technique directly and mainly influences two factors LS and P in addition to area exposed to erosion) as:

- The Support Practice Factor P: The support practice factor is difficult to define; it is not covered by this paper.
- The area exposed to rainfall and runoff is divided into meshes, the number of which depends on the project budget, limits of execution and amount of eroded soils permissible.

The slope length factor is a function of the lozenges geometry and especially the length of the diagonal that follows the slope. Choosing a lozenge that has unequal diagonals, such as lozenge's diagonal length that follows the slope is less than the perpendicular diagonal (Vertical diagonal's length = ½ horizontal diagonal's length), because:

- For the same area covered by the concrete lozenges, we have a shorter slope length.
- Concrete lozenges' perimeter is optimal.

The present study is based on the hypothesis that: the lozenges limit the formation of runoff such that the equivalent runoff length for each lozenge is the half of diagonal's length.

In selecting Wishmeier's model for predicted soil loss calculation down a slope, it is important to remember its limitations. Indeed, it does not take into account the interactions between factors.

References

1. Horton, R.E.: The rôle of infiltration in the hydrologic cycle (Voorheesville, New York) transactions. Am. Geophys. Union **14**(1), 446–460 (1933)
2. Ennadifi, Y.: Etude géologique du Prérif oriental et son avant-pays (région comprise entre Mezguitem, Ain Zora etTizroutine). Thèse de doctorat de troisième cycle en science de la terre, 100 p. (1972). https://tel.archives-ouvertes.fr/tel-00761774

3. Sadiki, A., Bouhlassa, S., Auajjar, J., et al.: Utilisation d'un SIG pour l'évaluation et la cartographie des risques d'érosion par l'equation universelle des pertes en sol dans le Rif oriental (Maroc): cas du bassin versant de l'oued Boussouab. Bulletin de l'Institut Scientifique, Rabat, Section Sciences de la Terre **2004**(26), 69–79 (2004)
4. Tribak, A., El Garouani, A., Abahrour, M.: L'érosion hydrique dans les séries marneuses tertiaires du prérif oriental: agents, processus et évaluation quantitative. Rev. Mar. Sci. Agron. Vét. **1**, 47–52 (2012)
5. Roose, É.: Introduction à la gestion conservatoire de l'eau, de la biomasse et de la fertilité des sols (GCES). Bull. Pédol. FAD **70**, 420 (1994)
6. Novec & Tanger Med Ingeneering: Etude d'impact sur l'environnement du nouveau port Nador West Med, mission 01: Etude d'Impact Envionnemental Novec, version définitive - N714-13c (2014)
7. Ministère des travaux publics et des communications direction de l'hydraulique division des ressources en eau ressources en eau du Maroc. Tome 1 domaines du rif et du Maroc oriental. Editions du service geologique (RABAT-Maroc) (1971)
8. Renard, K.G.G.R., Foster, G.A., Weesies, D.K., et al.: Coordinators. Predicting Soil Erosion by Water: A Guide to Conservation Planning With the Revised Universal Soil Loss Equation (RUSLE). USDA, United State Department of Agriculture, Agricultural Research Service Agriculture Handbook, no. 703, 404 pp. (1997)
9. Rango, A., Arnoldus, H.M.J.: Aménagement des bassins versants. Cahiers techniques de la FAO (1987)
10. Wischmeier, W.H., Smith, D.D.: Prediction rainfall erosion losses, a guide to conservation planning. Science U.S. Department Agriculture. Agriculture Handbook **537**, 60 p. (1978)
11. Chehlafi, A., Kchikach, A., Derradji, A.: Protection des talus autoroutiers par arcades bétonnées ou maçonnées, pp. 291–300. Rock Slope Stability, Marakech (Maroc) (2014)
12. Foster, G.R., Meyer, L.D., Onstad, C.A.: A runoff erosivity factor and variable slope length exponents for soil loss estimates. Trans. ASAE **20**, 683–687 (1977)
13. McCool, D.K., Brown, L.C., Foster, G.R., et al.: Revised slope steepness factor for the universal soil loss equation. Trans. ASAE **30**(1387), 1396 (1987)
14. Jean Monfet, ing. agr. Évaluation du coefficient de ruissellement à l'aide de la méthode SCS modifiée. Gouvernement du Québec Ministère des Richesses naturelle: Service de l'hydromètrie. Québec. HP-5 (1979)
15. Shin, G.J.: The analysis of soil erosion analysis in watershed using GIS. Ph.D. Dissertation, Department of Civil Engineering. Gang-won National University (1999)

Effect of Salt Water on Unconfined Compressive Strength for Cement Kiln Dust

Mahmoud E. Hassan[1(✉)], Ayman L. Fayed[2], and Mohamed Y. Abd El-Latif[2]

[1] Construction and Building, 6th University, Giza, Egypt
mahmoud_elsayed_151189@hotmail.Com
[2] Geotechnical Engineering Department, Ain Shams University, Cairo, Egypt

Abstract. CKD is a fine cement powder produced in large amounts as a by-product during cement manufacturing. It is mainly composed of oxidized, anhydrous and micron-sized particles. It is considered a major health hazard where the Ministry of Environment recommended its beneficial utilization uses or getting rid of it.

The maximum load, which can be transmitted to the soil by foundation, depends on the resistance of soil to the shearing deformation and compressibility. In this study, unconfined compression strength tests were performed on fresh CKD samples.

Different factors affecting the samples preparing process were investigated in the current experimental study including the amount of water required to form the samples and the type of the mixing water (fresh water and salt water). The percentages of water were 50%, 60% and 70% from CKD sample weight.

Due to the proximity of most of the cement factories in Egypt to the sea, the experimental study confirmed that using the salt water in unconfined compressive strength test for the CKD is the most favorable method. Also, the tests proved the ability to utilize the CKD rather than disposal in landfills. Results of the performed experimental study showed that, using of salt water/seawater leads to increase the unconfined compressive strength and decreasing the failure strain.

1 Introduction

Worldwide, cement manufacturing is a strategic industry. Likewise, it is the case in Egypt, as it is among the largest producers in the Middle East, Africa and the Arab Region. Egypt produces 70 million tons of clinker. Cement By-Pass Dust “CBPD” or cement Kiln Dust (CKD), is a fine sticky powder produced in large amounts as a by-product during cement manufacturing. It is composed of oxidized, anhydrous and micron-sized particles. It is considered as a health hazard, as the Ministry of Environment recommended utilizing it beneficially or to get rid of it.

Many researches are involved in investigating the utilization of CKD worldwide in order to find an economically efficient way for its recycling or its applications (i.e. soil stabilization, pavements, landfills and concrete mixes). However, according to the United States (Todres et al. 1992), more than four million tons are not suitable for

M. Shehata et al. (Eds.): GeoMEast 2019, SUCI, pp. 43–58, 2020.
https://doi.org/10.1007/978-3-030-34249-4_5

recycling. Accordingly, they must be disposed, annually. In this research, a real solution is provided to utilize the CKD by mixing it with salt to form samples can be used in various applications.

2 Materials

Different materials have been used during the scope of this work including the following:

2.1 Cement Kiln Dust

The main material was the cement Kiln Dust. Fresh CKD was brought from (Arabian Cement Company (ACC) in Ain Sokhna) on July 26, 2017. Arabian Cement Company (ACC) is the third largest cement producer in Egypt with a capacity of five million tons/year. Figure 1 shows the form of the utilized CKD sample.

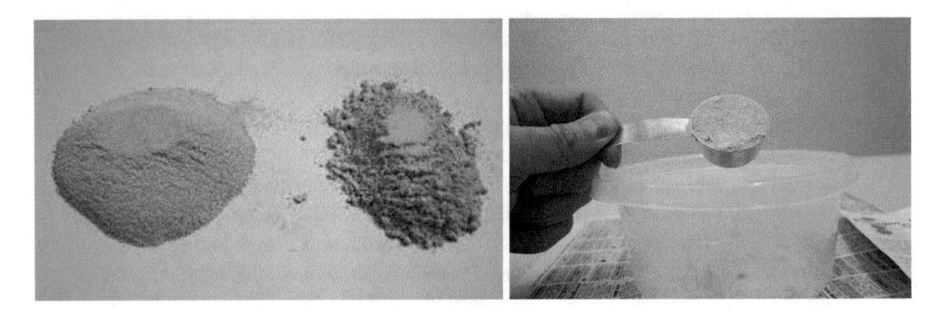

Fig. 1. Cement kiln dust sample (CKD)

2.2 Fresh Water

Fresh water utilized in the experimental work is the ordinary tap water.

2.3 Salt Water

Used salt water consisted of fresh tap water mixed with 35 parts per thousand of normal table salt, composed primarily of sodium chloride (NaCl). The used salt water is basically a simulation of the normal seawater with almost the same salinity.

3 Experimental Program

Details of the planned and executed experimental program are shown in Table 1. Initially, chemical and physical tests for the CKD samples such as the chemical analysis, grain size distribution and specific gravity tests were performed for characterization of the used material (CKD).

Table 1. Experimental program

Group no.	Test no.	Test type	Mixing fluid type	Test operating duration	Test sample
G1	1–1	Unconfined compressive strength	Fresh water	After 7 days	CKD + 50% W.C
	1–2		Salt water		
	1–3		Fresh water		CKD + 60% W.C
	1–4		Salt water		
	1–5		Fresh water		CKD + 70% W.C
	1–6		Salt water		

In this study, unconfined compression strength tests were performed on fresh CKD samples. The samples of CKD were prepared in manufactured mold with height of 6 inch and diameter of 3 inch to meet the aspect ratio requirement of 2:1 H/D. Load was applied on the samples with constant rate without any lateral support. It was increased until the failure occurs.

Fresh Cement Kiln Dust was used to prepare the sample of the test. Note that, 1 kg of the cement Kiln Dust was required to prepare one sample. Fresh water and salt water were used for mixing the cement Kiln Dust to prepare the samples. The water contents to form the samples were 50%, 60% and 70% of CKD weigh.

The results of the test for each percentage of water content were obtained from the average of three samples test. Note that, the samples were left for seven days to dry, where the target duration is 7 days, as the cement reaches approximately 70% of its strength after 7 days. All tests were performed according to the standards mentioned in Table 2.

Table 2. Specification and tests for the CKD samples

Type of Test	CKD
Chemical composition	ASTM C114
Particle size distribution	ASTM D422
Specific gravity	ASTM D-854-92
Unconfined compressive strength	ASTM D 2166-00

4 Test Results

4.1 Cement Kiln Dust Characterization Tests

4.1.1 Chemical Composition

Chemical analysis test was conducted in the Arabian Cement Company in Ain sokhna–Suez, using x-rays test by the winxrf equipment. The chemical composition was as given in Table 3 for the CKD sample by weight. The amount of the different chemical elements in the CKD can vary significantly from one plant to another depending on the raw materials and type of collection process, (Miller and Zaman 2003).

4.1.2 Particle Size Distribution

CKD

Particles size distribution test was conducted on the CKD samples using the (151-H) hydrometer in accordance with the standards shown in Table 2. The grain size distribution curve results for CKD sample is shown in Fig. 2 representing a generally fine material with almost 90% in the size of silt (0.002 > D > 0.06).

Table 3. Chemical composition for the CKD sample

Date	SiO_2	Al_2O_3	Fe_2O_3	CaO	MgO	SO_3	K_2O	Na_2O	Cl	LOI	SUM	LSF	SIM	ALM
26-Jul	13.857	4.528	2.986	56.909	2.966	2.18	5.515	1.488	7.459	2.112	99.00	123.4909	1.844158	1.51641

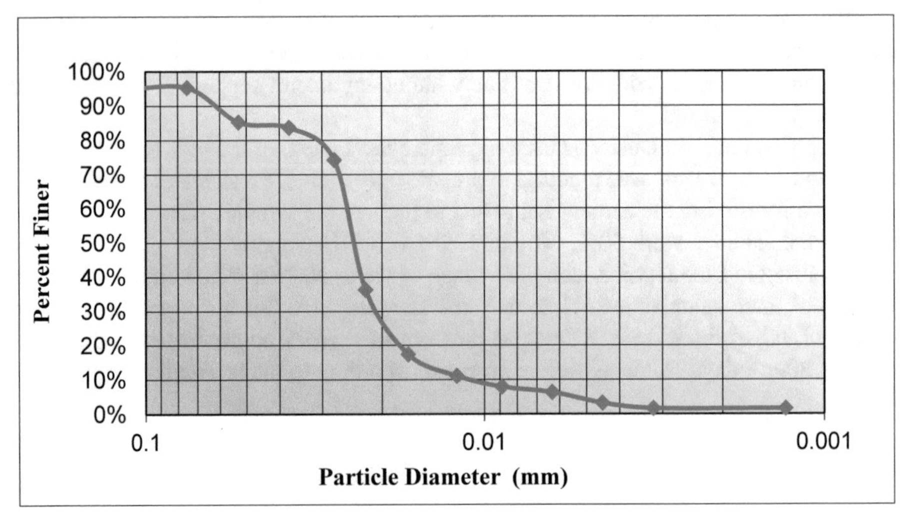

Fig. 2. Grain size distribution curve for CKD sample

Specific Gravity Test (Gs)

For CKD sample, the test was conducted on three CKD samples giving an average Gs value of approximately 2.738. The specific gravity of CKD is typically in the range of 2.6 to 2.8, (Baghdadi et al. 1995), which is usually less than that of the Portland cement (Gs ~ 3.15).

4.2 Unconfined Compressive Strength of CKD

Unconfined compressive strength test was conducted on CKD samples mixed with fresh and salt water at the certain water contents of 50%, 60% and 70%. The samples were left in the manufactured mold for one day. The samples were extracted from the mold and were dried for 7 days. Note that the results of each test were obtained from the average of 3 samples test.

Figure 3 presents samples of CKD after drying for 7 days before rejecting the bad samples. Figure 4 shows acceptation tested samples of CKD after drying for 7 days. The surfaces of the accepted tested samples were equalized by the electric saw as shown in Fig. 5. Note that, the samples that did not satisfy these measures were rejected and alternate samples were prepared. Figure 6 shows failure shape of acceptation tested sample.

Fig. 3. Samples of CKD after drying for 7 days before rejecting the bad samples

Fig. 4. Acceptation tested samples of CKD after drying for 7 days

Fig. 5. Samples with clean and equal surfaces

Fig. 6. Failure shapes of acceptation tested samples

Unconfined Compressive Strength for CKD Sample Mixed with 50% Fresh Water
Figure 7 presents the average stress/strain curve for the three samples of CKD mixed with 50% fresh water. The obtained compressive strength and its corresponding failure strain for the average of the three samples of CKD mixed with 50% fresh water are 6.35 kg/cm^2 and 1.35%, respectively.

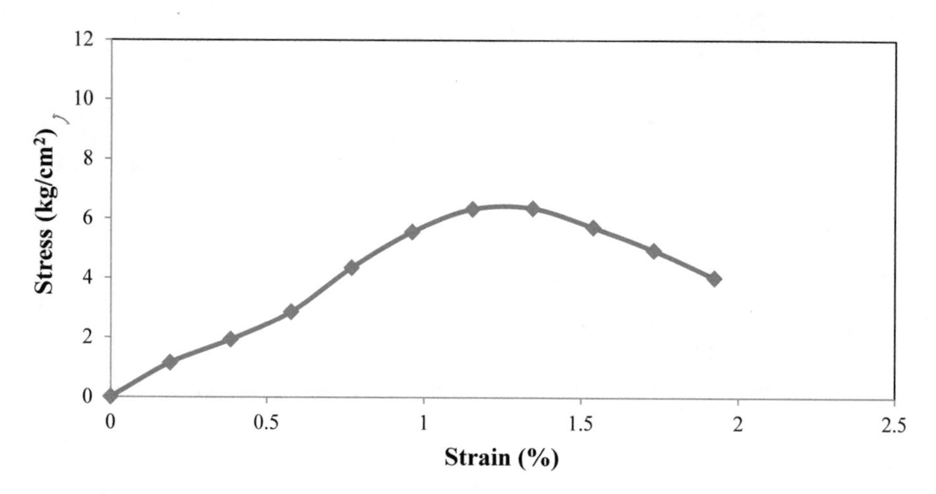

Fig. 7. Average stress/strain curve for the three samples of CKD mixed with 50% fresh water

Unconfined Compressive Strength for CKD Sample Mixed with 50% Salt Water

Figure 8 presents the average stress/strain curve for the three samples of CKD mixed with 50% salt water. The obtained compressive strength and its corresponding failure strain for the average of the three samples of CKD mixed with 50% salt water are 8.14 kg/cm^2 and 1.15%, respectively.

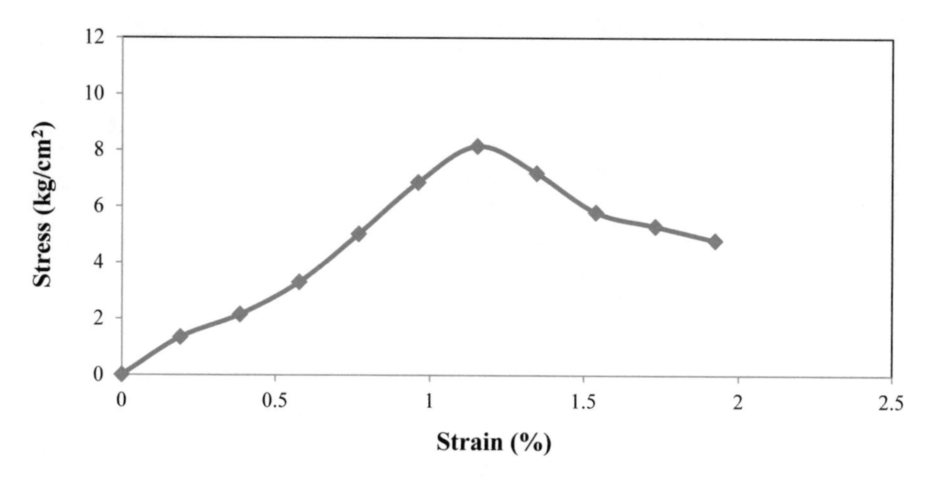

Fig. 8. Average stress/strain curve for the three samples of CKD mixed with 50% salt water

Unconfined Compressive Strength for CKD Sample Mixed with 60% Fresh Water

Figure 9 presents the average stress/strain curve for the three samples of CKD mixed with 60% fresh water. The obtained compressive strength and its corresponding failure strain for the average of the three samples of CKD mixed with 60% fresh water are 8.47 kg/cm^2 and 1.92%, respectively.

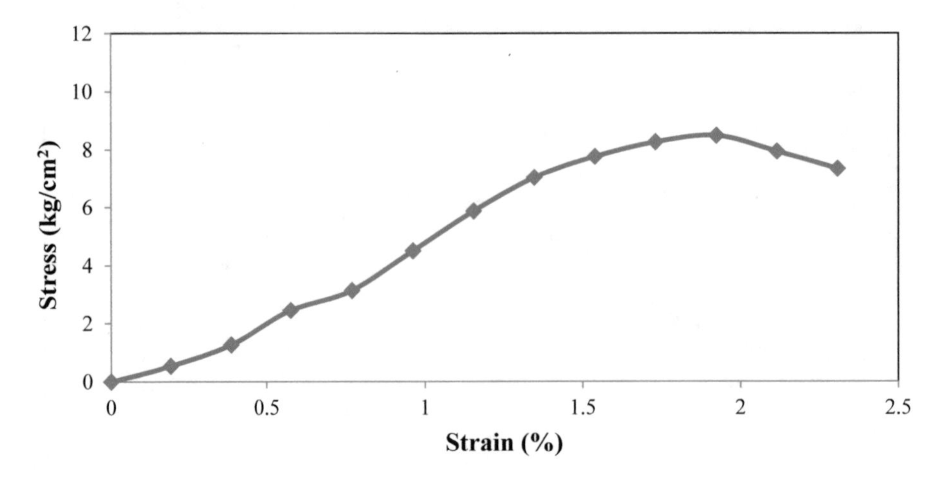

Fig. 9. Average stress/strain curve for the three samples of CKD mixed with 60% fresh water

Unconfined Compressive Strength for CKD Sample Mixed with 60% Salt Water
Figure 10 presents the average stress/strain curve for the three samples of CKD mixed with 60% salt water. The obtained compressive strength and its corresponding failure strain for the average of the three samples of CKD mixed with 60% salt water 9.88 kg/cm^2 and 1.73%, respectively.

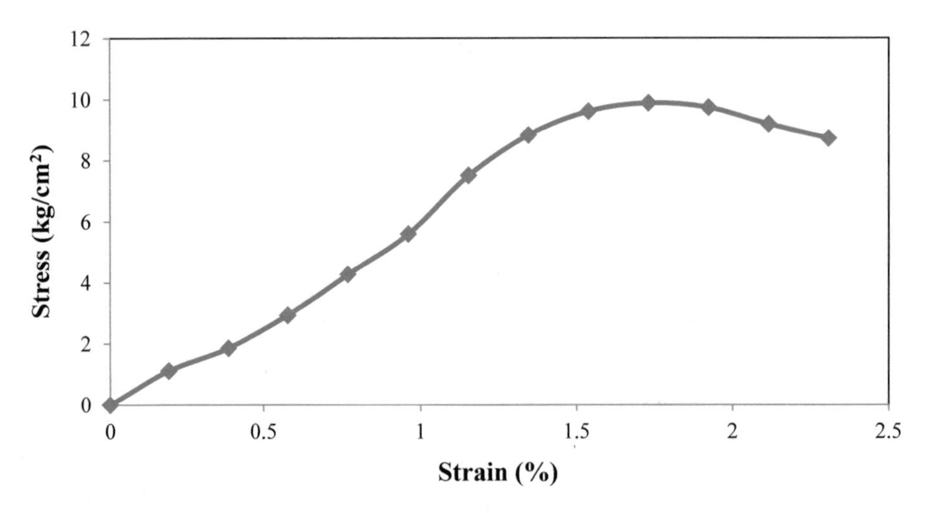

Fig. 10. Average stress/strain curve for the three samples of CKD mixed with 60% salt water

Unconfined Compressive Strength for CKD Sample Mixed with 70% Fresh Water
Figure 11 presents the average stress/strain curve for the three samples of CKD mixed with 70% fresh water. The obtained compressive strength and its corresponding failure strain for the average of the three samples of CKD mixed with 70% fresh water are 7.69 kg/cm^2 and 1.35%, respectively.

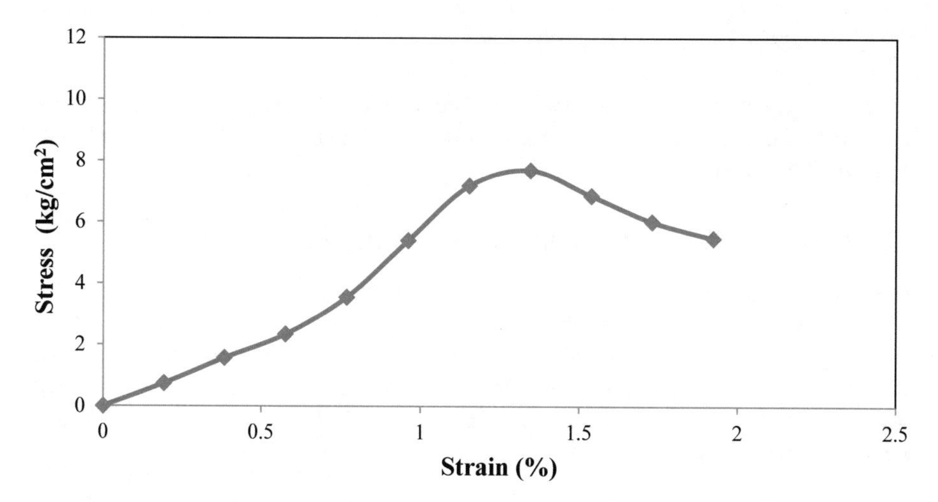

Fig. 11. Average stress/strain curve for the three samples of CKD mixed with 70% fresh water

Unconfined Compressive Strength for CKD Sample Mixed with 70% Salt Water

Figure 12 presents the average stress/strain curve for the three samples of CKD mixed with 70% salt water. The obtained compressive strength and its corresponding failure strain for the average of the three samples of CKD mixed with 70% salt water 8.20 kg/cm^2 and 0.96%, respectively.

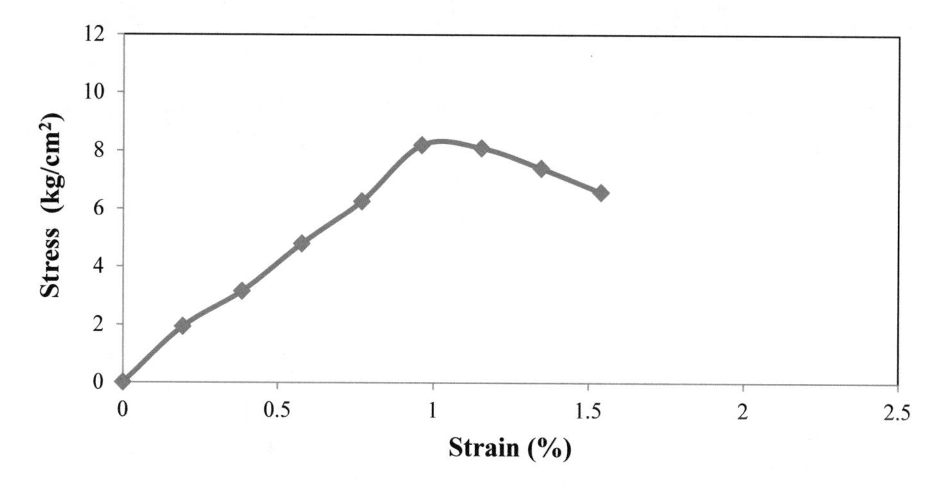

Fig. 12. Average stress/strain curve for the three samples of CKD mixed with 70% salt water

5 Comparisons and Discussions

Results of the performed tests are discussed in the following sections in order to understand the effect of different mixing ratios on UCS results and the effect of water type on UCS results.

5.1 Effect of Different Mixing Ratios on UCS Results

Table 4 presents the results of UCS tests for CKD samples mixed with 50%, 60% and 70% fresh water and salt water fresh water and salt water. Figures 13 and 14 related to Table 4 present the average stress/strain curve for the three samples of CKD mixed with 50%, 60% and 70% fresh water and salt water respectively. For the fresh water case, obvious is that the CKD sample, mixed with 60% fresh water, gave the highest value of compressive strength.

Table 4. Results of UCS tests for CKD samples mixed with 50%, 60% and 70% fresh water and salt water

Water type	Fresh water		Salt water	
Adding water percentage (%)	Max unconfined compressive stress (kg/cm^2)	Failure strain (%)	Max unconfined compressive stress (kg/cm^2)	Failure strain (%)
50	6.35	1.35	8.14	1.15
60	8.47	1.92	9.88	1.73
70	7.69	1.35	8.2	0.96

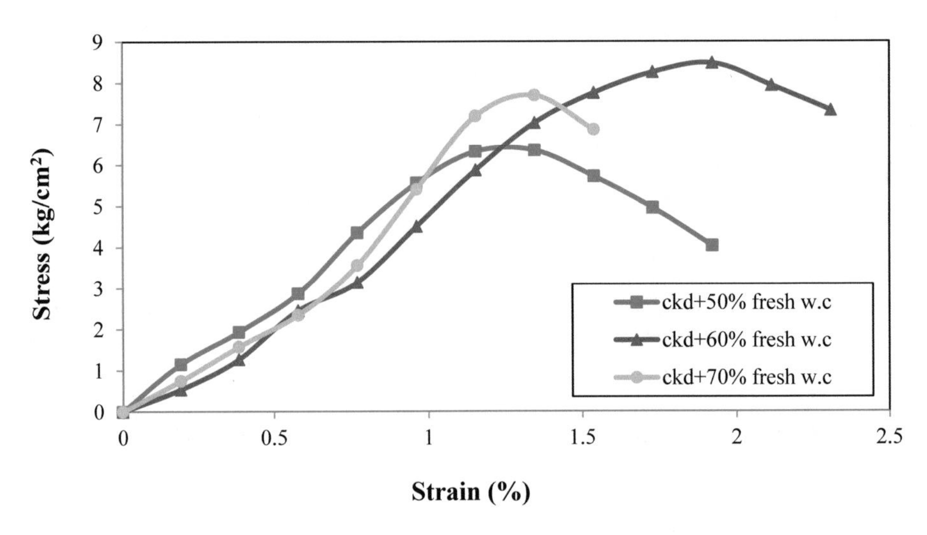

Fig. 13. Average stress/strain curve for the three samples of CKD mixed with 50%, 60% and 70% fresh water

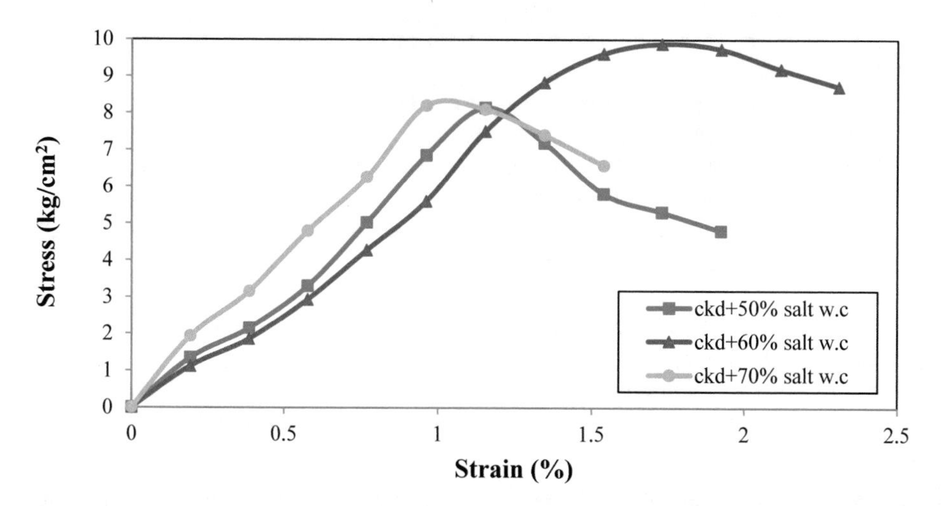

Fig. 14. Average stress/strain curve for the three samples of CKD mixed with 50%, 60% and 70% salt water

For the salt water case, noticeable is that the CKD sample, mixed with 60% salt water, gave the highest value of compressive strength.

Effect of Mixing Ratios on UCS Values for CKD Samples Mixed with Fresh and Salt Water

Figure 15 and Table 4 present the effect of mixing ratios on UCS values for CKD samples mixed with fresh and salt water, from the results, it can be concluded that:

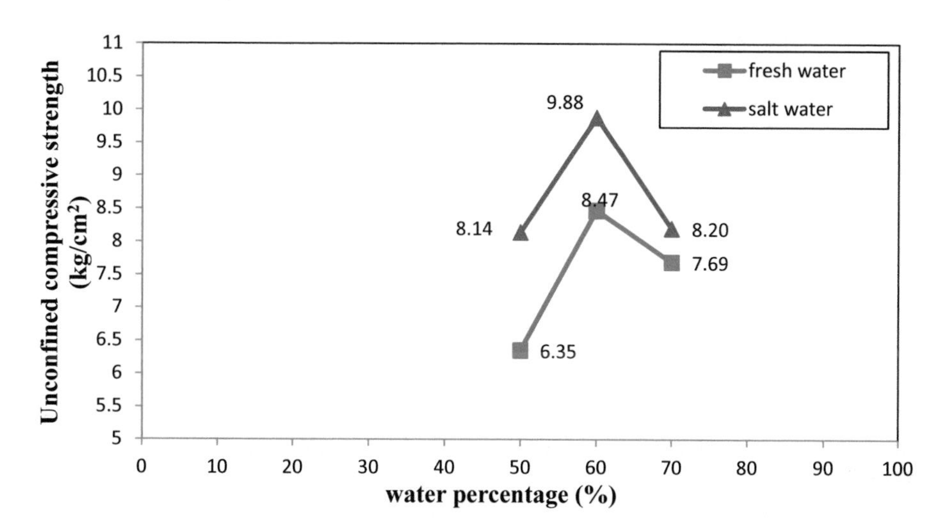

Fig. 15. Effect of mixing ratios on UCS values for CKD samples mixed with fresh and salt water

- For the fresh water, the 60% mixing ratio increased the UCS by about 25.03% (8.47 km/cm^2 versus 6.35 km/cm^2).
- For the fresh water, the 70% mixing ratio increased the UCS by about 17.43% (7.69 km/cm^2 versus 6.35 km/cm^2).
- For the salt water, the 60% mixing ratio increased the UCS by about 17.61% (9.88 km/cm^2 versus 8.14 km/cm^2).
- For the salt water, the 70% mixing ratio increased the UCS by about 0.73% (8.20 km/^{c}m^2 versus 8.14 km/^{c}m^2).

Effect of Mixing Ratios on the Failure Strain Values for CKD Samples Mixed with Fresh and Salt Water

Figure 16 and Table 4 present the effect of mixing ratios on the failure strain values for CKD samples mixed with fresh and salt water, from the results, it can be concluded that:

- For the fresh water, the 60% mixing ratio increased the failure strain by about 29.69% (1.92% versus 1.35%).
- For the fresh water, the 70% mixing ratio, the failure strain still the same (1.35%).
- For the salt water, the 60% mixing ratio increased the failure strain by about 33.53% (1.73% versus 1.15%).
- For the salt water, the 70% mixing ratio decreased the failure strain by about 16.52% (0.96% versus 1.15%).

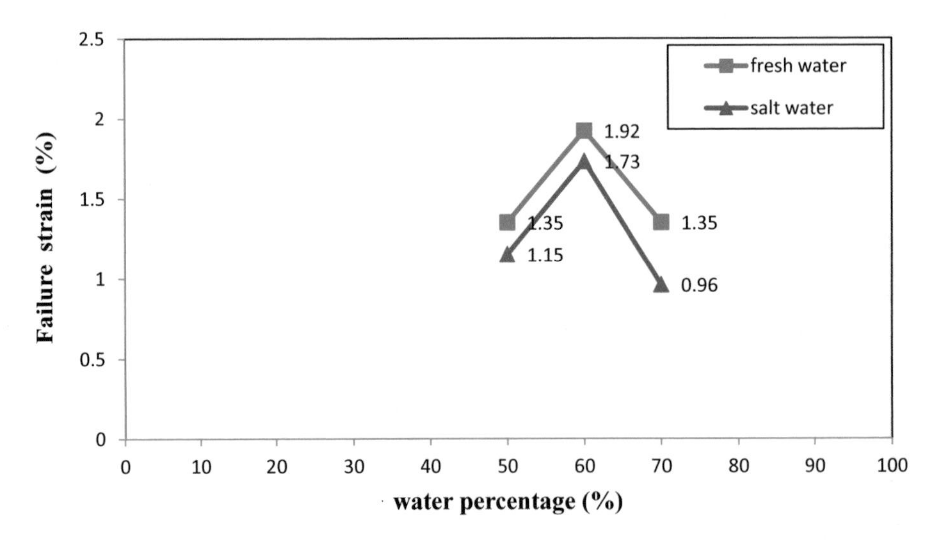

Fig. 16. Effect of mixing ratios on failure strain for CKD samples mixed with fresh and salt water

5.2 Effect of Water Type on UCS Results

The effect of water type on unconfined compressive stress and the effect of water type on failure strain (%) are discussed in the following sections.

Effect of Water Type on UCS Values for CKD Samples Mixed with Fresh and Salt Water

Figure 17 and Table 4 present the effect of water type on UCS values for CKD samples mixed with fresh and salt water, from the results, noticeable is that using salt water instead of fresh water led to increase the UCS. First for the 50% mixing ratio, using of salt water instead of fresh water led to increase the UCS by about 21.99%. From the results as well, for the 60% mixing ratio, the UCS increased by about 14.27%. Finally for the 70% mixing ratio, the UCS increased by about 6.22%.

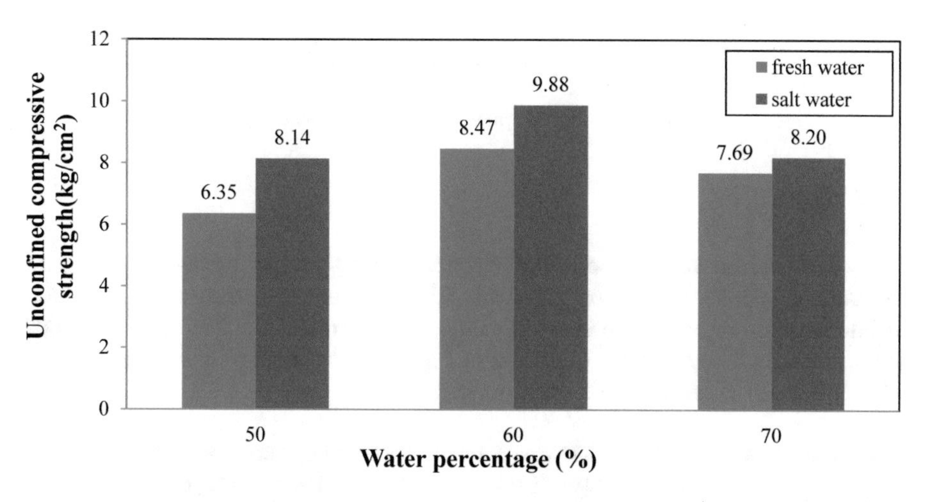

Fig. 17. Effect of water type on unconfined compressive stress

Effect of Water Type on the Failure Strain Values for CKD Samples Mixed with Fresh and Salt Water

Figure 18 and Table 4 present the effect of water type on the failure strain values for CKD samples mixed with fresh and salt water, from the results, noticeable is that using salt water instead of fresh water led to increase the UCS and decrease the failure strain. First for the 50% mixing ratio, using of salt water instead of fresh water led to decrease the failure strain by about 14.81%. From the results as well, for the 60% mixing ratio, the failure strain decreased by about 9.90%. Finally for the 70% mixing ratio, the failure strain decreased by about 28.89%.

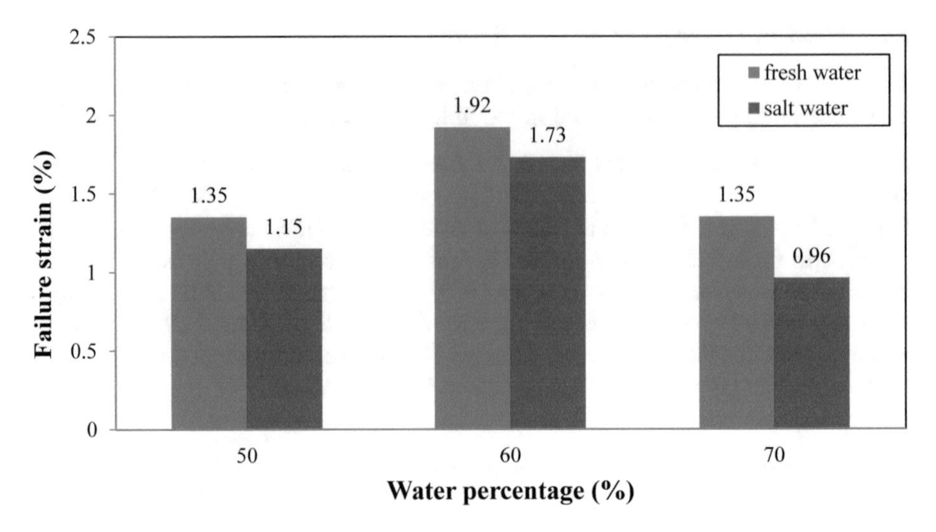

Fig. 18. Effect of water type on failure strain (%)

6 Conclusions

Salt water leads to an increase in the unconfined compressive strength and decreases the maximum strain. In addition, it increases the cohesion (C) and decreasing the angle of internal friction (Φ) in the shear stress test. Furthermore, CKD samples with 60% fresh water gave higher value of unconfined compressive strength than those samples with 50% fresh water and 70% fresh water.

Acknowledgments. The used CKD material and corresponding chemical analysis results are provided by Arabian Cement Company (ACC) in Ain Sokhna, Egypt. Their permission for using this material and provided information are gratefully acknowledged.

References

Abeln, D.L., Hastings, R.J., Scxhreiber, R.J., Yonley, C.: Detailed illustration of contingent management practices for cement Kiln dust. Research and Development Bulletin, SP115T, Portland Cement Association, Skokie, IL (1993)

American Society of Testing and Materials (ASTM): Standard Test Method for Chemical Analysis of Hydraulic Cement. Designation C 114-03. ASTM International, West Conshohocken

American Society of Testing and Materials (ASTM): Standard Test Method for Particle Size Analysis of Soils. Designation D 422-02. ASTM International, West Conshohocken

American Society of Testing and Materials (ASTM): Standard test method for sieve analysis of fine and ciarse aggregates C 136. Annual Book of ASTM Standards. ASTM International, West Conshohocken (199)

American Society of Testing and Materials (ASTM): Standard Test Method for Specific-Gravity-of-Solids Determination. Designation D 854-92. ASTM International, West Conshohocken

American Society of Testing and Materials (ASTM): D 2166-00. Standard test methods for unconfined compressive strength of cohesive soils. Annual Book of ASTM Standards. ASTM International, West Conshohocken

American Society of Testing and Materials (ASTM): Standard Guide for Evaluating Effectiveness of Admixtures for Soil Stabilization. Designation D 4609-01. ASTM International, West Conshohocken

Angelbeck, D., Burnham, J., Nicholson, J.P.: A new innovative sludge stabilization/management process: cement kiln dust (CKD) alkaline stabilization compared to anaerobic digestion-an economic analysis. Water Pollution Control Federation, December 1989

Baghdadi, Z.A., Fatani, N., Sabban, N.A.: Soil modification by cement kiln dust. J. Mater. Civ. Eng. ASCE **7**(4), 218–222 (1995)

Bhat, S.T., Lovell, C.W.: Use of coal combustion residues and waste foundry sands in flowable fill. Joint Highway Research Project. FHWA/IN/JHRP-96/2. Final Report (1997)

Bhatty, J.I.: Alternative uses of cement kiln dust. Research and Development Bulletin, RP327, Portland Cement Association, Skokie, IL, USA (1995)

Bhatty, J.I., Bhattacharja, S., Todres, H.A.: Use of cement kiln dust in stabilizing clay soils. Research & Development Bull, RP343, Portland Cement Association, Skokie, IL (1996)

Bye, G.C.: Portland Cement-Composition, Production and Properties. The Institute of Ceramics, Pergamon Press, USA (1983)

Collins, R.J., Emery, J.J.: Kiln dust-fly ash system for highway bases and subbases. Federal Highway Administration Report FHWA/RD-82/167. U.S Department of Transportation, Washington D.C (1983)

Gebhart, L.R.: Preliminary glossary of terms relating to grouting. J. Geotech. Eng. Divi. **106** (GT7), 803–815 (1980)

Diamond, S., Kinter, E.B.: Mechanism of soil-lime stabilization-an interpretive review. Highway Res. Rec. **92**, 83–102 (1965)

Du, L., Folliard, K.J., Trejo, D.: Effects of constituent materials and quantities on water demand and compressive strength of controlled low-strength material. J. Mater. Civ. Eng. **14**(6), 485–495 (2002)

Grindrod, P.S.: Application of the andreasen pipet to the determination of particle size distribution of portland cement and related materials. Fineness of Cement. ASTM STP 473, American Society for Testing Materials, pp. 45–70 (1970)

Haynes, W.B., Kramer, G.W.: Characterization of U.S. Cement Kiln Dust. Information Circular #8885, U.S Bureau of Mines, U.S. Department of the Interior, Washington D.C (1982)

Klemm, W.A.: Kiln dust utilization. Martin Marietta Laboratories Report MML, pp. 80–112, Baltimore (1980)

Kumar, R., Kanaujia, V.K., Ranjan, A.: An experimental study on potential cement kiln dust in stabilization of fly ash. Cem. Concr. Aggregates **24**(1), 25–27 (2002)

MacKay, M., Emery, J.: Stabilization and solidification of contaminated soils and sludges using cementitious system-selected case histories. Transp. Res. Rec. **1458**, 67–72 (1992)

McCoy, W.J., Kriner, R.W.: Use of Waste Kiln Dust for Soil Consolidation. Mill Session Papers, Portland Cement Association, Skokie (1971)

Miller, G.A., Zaman, M., Rahman, J., Tan, K.N.: Laboratory and field evaluation of soil stabilization using cement kiln dust. Final Report, No. ORA 125-5693, Planning and Research Division, Oklahoma Department of Transportation (2003)

Miller, G.A., Azad, S.: Influence of soil type on stabilization with cement kiln dust. Constr. Build. Mater. **14**(2), 89–97 (2000)

Muller, H.P.: What is Dust?: Characterization and classification of kiln dust. In: 24th Technical Meeting, Report No. MA 77/2505/E, Holderbank Management and Consulting Ltd., Technical Center, Material Division, Aargau (1977)

Nisbet, M.: The 3Rs and cement kiln dust: opportunities for reduction, reuse and recycling. For Presentation at the Air & Waste Management Association's 90th Annual meeting & Exhibition, Toronto (1997)

Peethamparan, S., Olek, J., Helfrich, K.E.: Evaluation of the engineering properties of cement kiln dust (CKD) modified kaolinite clay. In: The 21st International Conference on Solid Waste Technology Management, Philadelphia, 26–29 March 2006

Sayah, A.I.: Stabilization of expansive clay using cement kiln dust. M.S. thesis, University of Oklahoma, Norman (1993)

Todres, H.A., Mishulovich, A., Ahmed, J.: Cement kiln dust management: permeability. Research and Development Bulletin RD103T, Portland Cement Association, Skokie, Illinois, USA (1992)

Udoeyo, F.F., Hyee, A.: Strength of cement kiln dust concrete. J. Mater. Civ. Eng. ASCE **14**(6), 524–526 (2002)

Zaman, M., Laguros, J.G., Sayah, A.: Soil stabilization using cement kiln dust. In: Proceedings 7th International Conference on Expansive Soils, Dallas, pp. 347– 351 (1992)

Use of Hot Spring Bacteria for Remediation of Cracks in Concrete

Saroj Mandal[1(✉)] and B. D. Chattopadhyay[2]

[1] Civil Engineering Department, Jadavpur University, Kolkata, India
mailtosarojmandal@rediffmail.com
[2] Physics Department, Jadavpur University, Kolkata, India

Abstract. Despite its versatility in construction, concrete is known to have several limitations. It is weak in tension, poor in ductility and low in resistance to cracking. Its strength and durability suffer due to physical stresses, biogenic corrosion and chemical attacks. Cracks and fissures are typical signs in such situations. An attempt has been made to use a novel thermophilic hot spring bacteria to repair cracks and increase the durability in specimens of concrete/mortar. Mortar samples incorporating bacteria and cured in hot condition, showed a significant decrease in sulphate and water absorption capacities and a significant increase in compressive strength. Ultrasonic velocity test conducted on such samples confirmed more compactness in the sample. The bacteria incorporated concrete showed better performance in cracks repairing compared to the normal cement-sand mixture. The results clearly showed that remediation for cracks and durability of concrete structures can be enhanced with the addition of bacteria optimised at a cell concentration of 10^5 cells/ml water.

1 Introduction

Bacterial concrete is a new concept in the field of concrete technology and is self-repairing biomaterial that can remediate cracks and fissures in concrete through microbial processes. Though concrete is quite strong mechanically, it suffers from low tensile strength. Other draw backs include: permeability to liquid and consequent corrosion of reinforcement; susceptibility to chemical attack and ensuing low durability. Freeze/thaw action, chemical attack and alkali aggregate reactions are the common deteriorating mechanisms in concrete. But, corrosion of reinforcement is perhaps the main cause of deterioration in most countries. In additions to the above, the degradation of mortar/concrete by the acidophilic microorganisms has been well established. Because permeability of the concrete plays an important role in the durability based design, various innovative approaches relating to permeability have been proposed. In this context, among the recent advancements, the development of super plasticised concrete mixtures, which give a very high fluidity at relatively low water content, is noteworthy. Hardened concrete with low porosity, generally results in high strength and high durability. Along with the requirement of speed and durability of construction, now there is a third requirement of environmental friendliness of construction material that is becoming increasingly important. Therefore, the use of blast furnace slag, fly ash, silica fume etc. in concrete, is gaining momentum. In spite of

M. Shehata et al. (Eds.): GeoMEast 2019, SUCI, pp. 59–64, 2020.
https://doi.org/10.1007/978-3-030-34249-4_6

these changes, formation of cracks in concrete remains a common phenomenon. Without immediate and proper treatment, initial and fine cracks in concrete structure tend to increase with time, eventually requiring costly repair. There are a large numbers of products available commercially, for repairing cracks in concrete including structural epoxy, resins, epoxy mortar and synthetic mixtures. Selection of material for repairs therefore requires, a careful evaluation based on application technology, cost of material, time available for repair, environmental friendliness of the product etc. It has been reported that microbial mineral precipitation resulting from metabolic activities of microorganisms in concrete improves the overall behaviour of concrete. With this background, an attempt has been made to study the properties of concrete or mortar using microorganisms to repair cracks and to improve durability. A novel thermophilic anaerobic bacterium obtained from the hot spring of Bakreshwar in West Bengal, India was used for this study. Tests for durability and cracks repairs were performed on samples that had this microorganism in the cement sand mixture which was employed for making mortar or concrete specimens.

2 Experimental Program

2.1 Materials

Ordinary Portland Cement 43 grade (IS 8112: 1989) and standard Ennor sand (IS 650-1991) were used for samples preparation. Well-graded coarse aggregate with maximum size 10 mm was used for concrete samples. Standard mortar cubes (70.6 mm × 70.6 mm × 70.6 mm) and standard concrete cubes (100 mm × 100 mm × 100 mm) were cast as described in an earlier paper by the author. Cast specimens were subjected to water absorption capacity measurement and ultrasonic pulse velocity test9. Standard mortar cylinders (100 mm diameter and 50 mm height) were prepared from moulded cylinders (100 mm diameter and 200 mm height) for water absorption and sorptivity tests. To test the sulphate resistance of mortar and concrete samples, standard bar (25 cm × 25 cm × 250 cm) and standard prism (100 cm × 100 cm × 500 cm) were prepared respectively. Cement to sand ratio was kept as 1:3 and water to cement ratio was fixed at 0.4 for all the mortar samples. The concrete mixture used had cement, sand, aggregate in the ratio of 1 : 1.5 : 3 respectively by weight and the water to cement ratio was fixed at 0.48. All the samples were cured in water. Bacterial cells from a well-grown culture (cell concentration 10^8/ml) were added to the mortar and concrete mixture with proper dilution as required through the mixing water. For each experiment, bacteria having three different cell concentrations (10^4/ml, 10^5/ml and 10^6/ml) were mixed with the samples. For each cell concentration, three samples were tested at a time and every set of experiment was repeated at least 4 times to get an average data.

2.2 Tests

Mortar cubes after casting with and without bacterial cells were cured in a hot water bath maintained at 65 °C. After curing, the compressive strength of the control and bacteria treated samples were measured. The velocity of propagation of compressional wave in mortar/concrete samples cured in water at room temperature or cured in water at 65 °C, was tested in Pundit plus PC1007 Ultrasonic pulse velocity meter as per ASTM C597-0213. From the measured ultrasonic Pulse velocity, the dynamic modulus of elasticity was calculated. This test was performed as per ASTM C1585-04 on the cylindrical mortar samples cured in water for 45 days. After curing, the samples were dried at 52 °C in an oven for 72 h. The mass and diameter of the samples were recorded. Then the side surfaces of each sample were sealed with a synthetic paint. The mass of the painted samples was accurately measured again. Next, samples were placed in a water bath such that 1 to 3 mm height of the samples was in water. The increase in mass due to water absorption through the bottom surface was recorded after 4 h.

Artificial cracks were created on the top surface in the middle of the both mortar cubes (crack size - 5 cm length × 0.3 cm thickness × 1.5 cm depth) and prismatic mortar bars (crack size - 3.5 cm length × 0.3 cm thickness × 1 cm depth) by inserting a metallic steel strip during casting, and removing it the next day. The samples were cured in water for 7 days. The cracks formed were then filled up either with normal cement-sand mixture or with bacteria incorporated (cell concentration 10^5/ml of water used) cement-sand mixture. After repairing the cracks in this manner, the samples were cured in water for 28 days. The compressive strength of the mortar cubes and flexural strength of the prismatic bars were then measured.

3 Results and Discussion

As stated earlier the purpose of this study was to observe the effect of the hot spring microorganism on the cracks remediation and the durability of concrete and mortar, because durability is an important factor in the service life of concrete structures. Data are given in the tables as mean of 12 estimations. Table 1 shows the development of compressive strength of mortar cubes prepared by varying concentrations of bacterial cells, The samples were cured at 65 °C and the strength measurement were taken at 3, 7 and 28 days. It was seen that the overall compressive strength of the mortar cubes increased with the addition of the anaerobic microorganism at all ages, compare to control specimen (without microorganism). The maximum increment in compressive strength was at the bacterial concentration of 10^5 cells. The improvement in strength was due to the deposition as well as uniform distribution of silicate phases by the bacteria present in increase the compactness in the samples.

Table 1. Compressive strength of mortar with ages at hot water (65 °C) curing

Cell concentration/ml of water	Cube compressive strength, MPa					
	3 days		7 days		28 days	
	Strength ± S.D	% Increase relative to control	Strength ± S.D	% Increase relative to control	Strength ± S.D	% Increase relative to control
Nil (control)	15.97 ± 0.05	–	32.31 ± 0.28	–	47.73 ± 0.08	–
10^4	18.31 ± 0.10	14.65	36.54 ± 0.27	13.09	54.37 ± 0.29	13.91
10^5	19.97 ± 0.22	25.05	39.97 ± 0.15	23.71	58.28 ± 0.09	22.10
10^6	17.37 ± 0.30	8.77	35.10 ± 0.96	8.64	49.96 ± 0.43	7.48

Ultrasonic pulse velocity tests were conducted to seek confirmation about compactness in control and bacterial mortars cured at normal and hot curing condition. The results shown in Tables 2 and 3 confirmed that compactness in bacterial mortar samples was more compared to non-bacterial mortar samples. Bacterial action explained earlier, resulted in increasing the density of mortar/concrete and this densification increased the ultrasonic pulse velocity through mortar/concrete and the dynamic modules of elasticity. The maximum improvement occurred at the bacteria cell concentration of 10^5 cells/ml.

Table 2. Ultrasonic pulse velocity in mortar cube with normal water curing

Cell concentration/ml of water	Ultrasonic pulse velocity of mortar in normal water curing (km/sec)		
	3 days	7 days	28 days
Nil	3.446	3.595	3.944
10^4	3.488	3.681	3.971
10^5	3.516	3.687	4.002
10^6	3.497	3.649	3.955

Table 3. Ultrasonic pulse velocity in mortar cube with hot water curing (65 °C)

Cell concentration/ml of water	Ultrasonic pulse velocity of mortar in normal water curing (km/sec)		
	3 days	7 days	28 days
Nil	3.493	3.639	3.960
10^4	3.505	3.696	4.007
10^5	3.524	3.707	4.016
10^6	3.497	3.683	3.978

It is clear that the addition of microorganism reduces the sorptivity. The minimum sorptivity of the mortar and concrete were found to be 0.181 mm/min$^{0.5}$ (Table 4) and 0.155 mm/min$^{0.5}$ (Table 5) respectively at a cell concentration of 10^5 cells/ml. This indicated that permeability of the bacteria added specimen decreases due to the modification of pore structure.

Table 4. Sorptivity of mortar cube

Cell concentration per ml	Initial mass gm	Mass after 4 h gm	Increase in mass gm	Area mm^2	Mass/area $\times 10^{-3}$ gm/mm^2	Soptivity $mm/mm^{0.5}$
Nil	722.00	743.33	21.33	5000	4.27	0.275
10^4	742.00	762.00	20.00	5000	4.00	0.258
10^5	722.01	736.01	14.00	5000	2.80	0.181
10^6	726.00	743.33	17.33	5000	3.47	0.224

Table 5. Sorptivity of concrete cube

Cell concentration per ml	Initial mass gm	Mass after 4 h gm	Increase in mass gm	Area	Mass/area $\times 10^{-3}$ gm/mm^2	Soptivity $mm/mm^{0.5}$
Nil	2560	2590	30	10000	3.0	0.194
10^4	2416	2444	28	10000	2.8	0.181
10^5	2440	2454	24	10000	2.4	0.155
10^6	2444	2470	26	10000	2.6	0.168

Table 6 shows the result of crack repairing in mortar cubes and prismatic bars. It was seen that cement sand mixed with bacteria repaired cracks more effectively than the normal cement-sand mixture. The bioremediation property of the hot spring bacteria was thus demonstrated.

Table 6. Results of crack repairing of mortar sample by addition of hot spring bacteria at 10^5 cells/ml of water used

Sample	Flexural strength of mortar prismatic bar MPa		Compressive strength of mortar cube MPa	
	Strength ± S.D	% Decrease relative to uncracked	Strength ± S.D	% Decrease relative to uncracked
Uncracked	4.982 ± 0.062	–	43.40 ± 0.31	–
Normal repairing	3.126 ± 0.039	37.25	36.11 ± 0.74	16.80
Bacterial repairing	3.703 ± 0.018	25.67	41.22 ± 0.82	5.02

4 Conclusions

From the present experimental studies, it can be inferred that the incorporation of the themophilic anaerobic bacterium isolated from hot spring results in the improvement of compressive strength both in mortar and concrete specimens. By utilising these bacteria in mortar/concrete, the durability of these materials can be increased. In addition, the bacteria have a potential for crack repairing. Therefore the novel hot spring bacteria can play a new role in modern concrete technology.

References

1. Gautam, D., et al.: Common structural and construction deficiencies of Nepalese. Innov. Infrastruct. Solut. J. (2016). https://doi.org/10.1007/s41062-016-0001-3
2. Ramachandran, S.K., Ramakrishnan, V., Bang, S.S.: Remeditation of concrete using microorganisms. ACI Mater. J. **98**, 3–9 (2001)
3. Mehta, P.K.: Advancement in concrete technology. Concr. Int. **21**(6), 69–75 (1999)
4. Ramachandran, V.S.: Concrete Admixtures Handbook: Properties, Science and Technology, p. 629. Noyes Publications, New Jersey (1984)
5. Diercks, M., Sand, W., Bock, E.: Microbial corrosion of concrete, cellular and molecular life science, 514–516 (2005)
6. Glassgold, I.L.: Shortcrete durability: an evaluation. Concr. Int. Des. Constr. **11**(8), 78–85 (1989)
7. Hilsdorf, H.K.: Durability of concrete – a measurable quantity? IABSE Rep. **57**(1), 111–123 (1989)
8. Norman, M.P.L., Beaudoin, J.J.: Mechanical properties of high performance cement binders reinforced with wollastonite micro-fibres. Cem. Concr. Res. 22981–22989 (1992)
9. Ramakrishnan, V., Panchalan, R.K., Bang, S.S.: Bacterial concrete. In: Proceedings SPIE, vol. 4234, p. 168 (2001)
10. Ghosh, P., Mandal, S., Chattopadhyay, B.D., Pal, S.: Use of microorganisms to improve the strength of cement mortar. Cem. Concr. Res. **35**(10), 1980–1983 (2005)
11. Ghosh, P., Mandal, S., Chattopadhyay, B.D.: Effect of addition of microorganism on the strength of concrete. The Indian Concr. J. **80**(4), 45–48 (2006)
12. Sarkar, M., et al.: Development of an improved E. coli bacterial strain for green and sustainable concrete technology. RSC Adv. **5**(41), 32175–32182 (2015)
13. Sarkar, M., et al.: Autonomous bioremediation of a microbial protein (bioremediase) in Pozzolana cementitious composite. J. Mater. Sci. **49**(13), 4461–4468 (2016)
14. Specification for 43-grade ordinary portland cement, IS 8112: Bureau of Indian Standard, New Delhi (1999)
15. Specification for Ennor sand, IS 650: Bureau of Indian Standard, New Delhi (1991)
16. Standard test method for pulse velocity through concrete, ASTM C597-02

A Comparative Study Between the Individual, Dual and Triple Addition of (S.F.), (T.G.P.) and (P.V.A.) for Improving Local Gypsum (Juss) Properties

Ahmed S. D. AL-Ridha[1(✉)], Ali A. Abbood[1], Ali F. Atshan[2], Hussein H. Hussein[3], Layth Sahib Dheyab[4], Mohammed Sabah Mohialdeen[5], and Hameed Zaier Ali[5]

[1] Structural Engineering, Department of Civil Engineering, College of Engineering, Mustansiriyah University, Baghdad, Iraq
ahmedsahibdiab@yahoo.com
[2] Structural Engineering, Department of Water Resources Engineering, College of Engineering, Mustansiriyah University, Baghdad, Iraq
[3] Departments Petroleum Engineering, College of Engineering, Baghdad University, Baghdad, Iraq
[4] Civil Engineering, Consultant Engineer, Baghdad, Iraq
[5] Department of Civil Engineering, College of Engineering, Mustansiriyah University, Baghdad, Iraq

Abstract. In this research, an attempt has been made to implement a comparative study concerning the effect of three additives (namely; S.F., T.G.P. and P.V.A.) on two essential properties of the local gypsum (Juss) which are (Compressive Strength and Setting Time) for the purpose of improving both of them (together).

This objective was carried out through the individual addition of each additive firstly, and the dual addition secondly (adding each one with and without the others), and the triple addition of all of them together.

The work plan involves casting eight Juss mixes according to the weighted contents of these additives (1.2% for S.F., 0.4% for T.G.P. and 4.0% for P.V.A.) individually, dually and altogether in addition to the reference mix (the mix of no additives). The (water/ Juss) ratio used for all above mentioned mixes was fixed to (0.3).

It was found that the individual addition of each additive improves one of the above mentioned properties and deteriorate the other. While the dual addition was found either to slightly improve one property or considerably improve one of them and reduce the deterioration in the other.

Finally it was found that the triple usage of all these additives (altogether) leads to the best possible improvement for both properties simultaneously.

Keywords: Local gypsum (Juss) · S.F. · T.G.P. · P.V.A. · Compressive strength · Setting time

M. Shehata et al. (Eds.): GeoMEast 2019, SUCI, pp. 65–79, 2020.
https://doi.org/10.1007/978-3-030-34249-4_7

1 Introduction

In the recent years gypsum products have been exceedingly used as in-door finishings. Houses, especially in the U.S.A. and Europa, are either built from or lined with gypsums-based products favored by architects because of their superior properties, such as obtainable availability of in-expensive raw materials, volumetric stability, acoustic and thermals insulations, fires resistance, quite lows toxicity and the comparatively low energy and temperature needed in its manufacture [1]. Gypsum is also used in several applications beyond the construction field: e.g. in making molds for ceramics product [2], in medicals [3], and dentals accessories or implants [4], furthermore, it is the major constituent in Portland cement in order to delay its setting time [5]. The numerous applications of gypsum plaster are primarily based on its specified properties [6, 7].

Many researchers have attempted to develop plaster characteristics and extend its range of applications through the addition of other materials [7–9]. One of these additives is "Silica gel" (a highly porous forms of silica), it is a by-products of the sodium silicates industry with fabulous heat and fire resistance, chemical-stability, along with a large specifics surfaces area, and high waters sensitivity. In addition, its erratic nature reduce density as well as thermals conductivity and promotes the high temperature durability of plaster composites with trivial loss of compressive strength [2, 10]. The yield strength, elastic modulus, and interior bond of plasterboards have been observed to increase when nano-SiO2 is added [11]. Silica fumes, in turn, are a very good pozzolanic material with high reactions rates, although it is rarely used with gypsums [12]. Many authors have reported that the additions of ultras-fine sand (U.F.S.) or micros-silica improves the mechanical properties of Portland cement pastes [13, 14].

The water/gypsum ratios have an influence on the basic physical characteristic of the hardened gypsums, such as its volumetric densities, totals open porosity, and other related characteristic such as it is moistures mechanicals thermals and acoustic insulation properties. The theoretical water-gypsum ratios necessary for the hydrations of calcium sulphate hemi-hydrate CaSO4·½H2O into calcium sulphate dehydrate CaSO4·2H2O = 0.187. Additional water, in a so-called over-stoicheiometric quantity, is necessary for the process of hardening of the gypsum paste. The properties of the hardened gypsum made from a gypsum paste by casting, pressing, or vibrating, depend on the values of the water-gypsum ratio [15].

2 Experimental Work

2.1 Materials

Gypsum

- Gypsum products

Materials that are resulted from the calcinations of gypsum (CaSO4.2H2O) and having the chemical composition of hemihydrate (CaSO4.1/2H2O) are called "Gypsum Products". Although they are identical in compositions and x-rays diffraction peaks, they are different in their physio- mechanical properties. They consist of three main types: locals juss, plasters, and dental stone, each type has several varieties [16]. The first type has our concern in this research.

- Local Gypsum (Juss)

The word "juss" is derives from the Assyrians word "jasso". Local juss in Iraq is a materials produced from calcined gypsums by the "Koor method" Gypsums rocks pieced are placed on opening in the koor domes and the heated source is at the base of the domes. Heating continue for (24) hours. The final products the juss is a mechanicals mixture of anhydrites, bassanite and un-burnt gypsum.

The local gypsum (juss) used as a main matrix in this project was calcium sulfate hemihydrate gypsum (CaSO4.½H2O), which was obtained from local market in Baghdad.

Silica Fume (S.F.)
Silica fumes are highly reactive pozzolanic substances and are a byproduct from the production of silicone or Ferro-silicon metals. It is a very fine powder and composed from the flue gases from electric is furnace. The silica fume that is used in this research is a product from Sica Manufacturer in Egypt and have the product name "Sika Fume-HR".

Tree Glue Powder (T.G.P.)
Tree glue is taken from trees called (Arak) usually grow in Iran, it is grinded and used primarily for wooden works, but here it is used (may be for the first time) as an additive to gypsum plaster mixes.

Polyvinyl Acetate (P.V.A.)
Polyvinyl acetate (P.V.A.), commonly referred to as wood glue, white glue, carpenters glue, school glue, Elmer's glue in the U.S., or PVA glue, is an aliphatic rubbery synthetic polymer with the formula ($C_4H_6O_2$), it is also used (unprecedentedly) here in our present work as an additive to gypsum works.

Mixing Water
Ordinary potable water was used for mixing to all gypsum mixes in this study.

2.2 Gypsum Mixes

Eight mixes of gypsum have been studied in this research according to their additives contents, precisely: S.F. content (by weight) 0.0% & 1.2%, T.G.P. content (by weight) 0.0% & 0.4% and P.V.A. contents (by weight) 0.0% and 4.0%. The (water/ juss) ratio used for all the above mentioned mixes was (0.3). The details of these mixes are listed in Table 1, and the diagram of their work plan is illustrated in shown in Fig. 1 (Diagram 1).

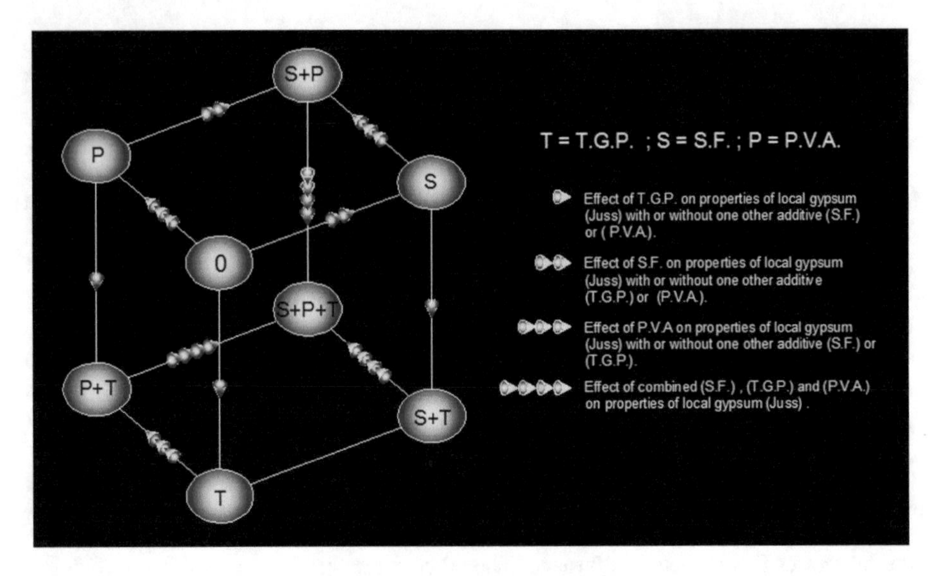

Diagram 1. Work plan Skeleton.

Table 1. Description of mixes.

Mix No.	S.F. content by weight (%)	P.V.A. content by weight (%)	T.G.P. content by weight (%)	(W/J) ratio	Ingredients Per (100 g) of Juss
Mix 1	0.0	0.0	0.0	0.3	(100 g) Juss + (0.0 g) S.F. + (0.0 g) P.V.A + (0.0 g)T.G.P. + (30 g) water
Mix 2	1.2	0.0	0.0	0.3	(100 g) Juss + (1.2 g) S.F. + (0.0 g) P.V.A + (0.0 g)T.G.P. + (30 g) water
Mix 3	0.0	4.0	0.0	0.3	(100 g) Juss + (0.0 g) S.F. + (4.0 g) P.V.A + (0.0 g)T.G.P. + (30 g) water
Mix 4	0.0	0.0	0.4	0.3	(100 g) Juss + (0.0 g) S.F. + (0.0 g) P.V.A + (0.4 g)T.G.P. + (30 g) water
Mix 5	1.2	4.0	0.0	0.3	(100 g) Juss + (1.2 g) S.F. + (4.0 g) P.V.A + (0.0 g)T.G.P. + (30 g) water
Mix 6	1.2	0.0	0.4	0.3	(100 g) Juss + (1.2 g) S.F. + (0.0 g) P.V.A + (0.4 g)T.G.P. + (30 g) water
Mix 7	0.0	4.0	0.4	0.3	(100 g) Juss + (0.0 g) S.F. + (4.0 g) P.V.A + (0.4 g)T.G.P. + (30 g) water
Mix 8	1.2	4.0	0.4	0.3	(100 g) Juss + (1.2 g) S.F. + (4.0 g) P.V.A + (0.4 g)T.G.P. + (30 g) water

2.3 Mixing Procedure

All mixes were made by weighted quantities (gypsum, S.F., T.G.P., P.V.A. and water). In the beginning S. F. and T.G.P. was added to the gypsum and be dry-mixed, then the specified quantity of the water was added to the mix, and re-mix manually for (approximately 30 s), then poured in to the mold. For mixes with P.V.A. additive, the required quantity of P.V.A. is added to the water, mixed carefully and then added to the dry mix. The mold has been vibrated benefiting from the vibration of a (small generator) for about 10 s. After 30 min, the cubic (5 × 5 × 5 cm) specimens were taken off from the mold. Then, the specimens were exposed to the direct sun light for about two days at approximately 38 °C heat.

2.4 Testing Program

Compression Strength

In this research, the 50 mm cubic specimens were tested at age of about one week or over to evaluate the compressive strength. Figure 1-a shows the testing machine used in our research [test is carried out according to ASTM: C472] [17].

Setting Time

One of the most disadvantages of gypsum mixes, precisely in the preparation of the gypsum paste is that its setting time is rather small (e.g. compared with cement or concrete paste) and this disadvantage doesn't provide enough comfort for the workers to do their job freely, this promotes us to investigate the effect of our additives (P.V.A., S.F. and T.G.P.) individually, two of them and all together on juss setting time.

Setting time is usually measured by a device called (Vicat apparatus), which consist of a 300 gm weighted rod ended with a needle (5 cm long) and (1 mm in diameter) fixed by a holder with a graduated plate and a semi-cone pan (60 * 70 * 40)mm dimensions, the apparatus is shown in Fig. 1-b, [test is carried out according to ASTM: C472-99] [17].

(a)

(b)

Fig. 1. (a): Compressive Strength Machine. (b): Vicat Apparatus

3 Results and Discussions

3.1 Compressive Strength

Effect of (S.F.) on Compressive Strength of Juss with and Without One Other Additive (T.G.P.) or (P.V.A.)

Figure 2 and Table 2 shows the effect of (S.F.) on the compressive strength of the juss. They illustrate that the compressive strength is increased by (33.8%) when (S.F.) is added alone to the mix, while the percentage of increasing becomes (21.31%) when (S.F.) is added together with (T.G.P.), and becomes (42.85%) when (S.F.) is added together with (P.V.A.).

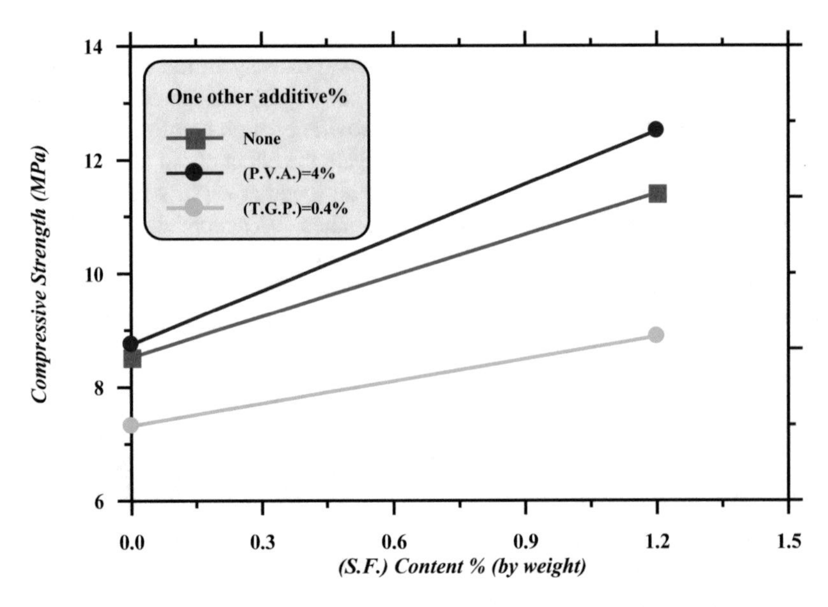

Fig. 2. Effect of (S.F.) on Compressive Strength of Juss with and without one other additive (T.G.P.) or (P.V.A.).

Table 2. Effect of (S.F.) on Compressive Strength of Juss with and without one other additive (T.G.P.) or (P.V.A.).

Mix No.	Content of one other additive by weight (%)	(S.F.) content by weight (%)	(W/J) ratio	Compressive strength (MPa)	Percentage of increase (%)
Mix 1	No other additive	0.0	0.3	8.52	——
Mix 2		1.2	0.3	11.4	33.8
Mix 4	(T.G.P.) content = 0.4%	0.0	0.3	7.32	——
Mix 6		1.2	0.3	8.88	21.31
Mix 3	(P.V.A.) content = 4%	0.0	0.3	8.75	——
Mix 5		1.2	0.3	12.5	42.85

Effect of (T.G.P) on Compressive Strength of Juss with and Without One Other Additive (S.F.) or (P.V.A.)

Figure 3 and Table 3 shows the effect of (T.G.P.) on the compressive strength of the juss. They illustrate that the compressive strength is increased by (14.08%) when (T.G.P.) is added alone to the mix, while the percentage of increasing becomes (22.1%) when (T.G.P.) is added together with (S.F.), and becomes (7.65%) when (T.G.P.) is added together with (P.V.A.).

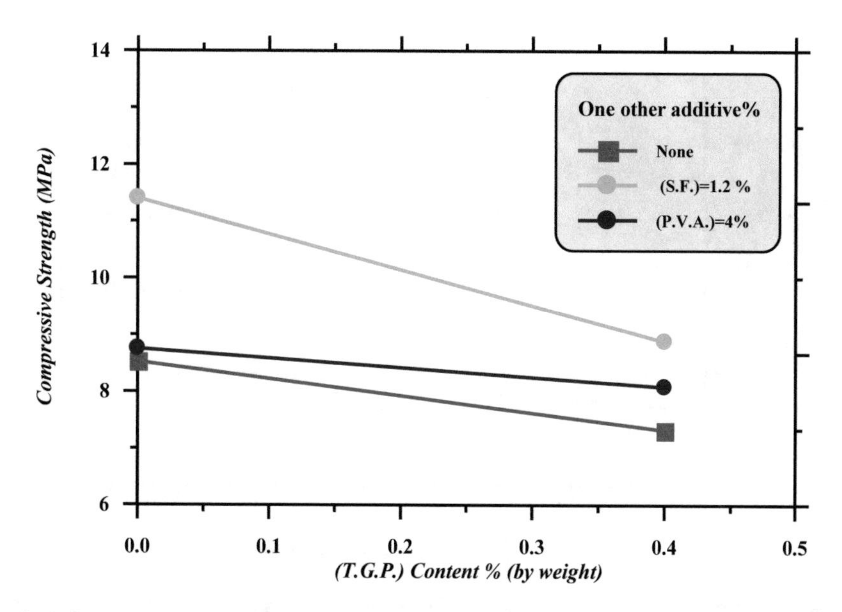

Fig. 3. Effect of (T.G.P.) on Compressive Strength of Juss with and without one other additive (S.F.) or (P.V.A.).

Table 3. Effect of (T.G.P.) on Compressive Strength of Juss with and without one other additive (S.F.) or (P.V.A.).

Mix No.	Content of one other additive by weight (%)	(T.G.P.) content by weight (%)	(W/J) ratio	Compressive strength (MPa)	Percentage of decrease (%)
Mix 1	No other additive	0.0	0.3	8.52	——
Mix 4		0.4	0.3	7.32	14.08
Mix 2	(S.F.) content = 1.2%	0.0	0.3	11.4	——
Mix 6		0.4	0.3	8.88	22.10
Mix 3	(P.V.A.) content = 4.0%	0.0	0.3	8.75	——
Mix 7		0.4	0.3	8.08	7.65

Effect of (P.V.A.) on Compressive Strength of Juss with and Without One Other Additive (S.F.) or (T.G.P.)

Figure 4 and Table 4 shows the effect of (P.V.A.) on the compressive strength of the juss. They illustrate that the compressive strength is increased by (2.69%) when (P.V.A.) is added alone to the mix, while the percentage of increasing becomes (9.64%) when (P.V.A.) is added together with (S.F.), and becomes (10.38%) when (P.V.A.) is added together with (T.G.P.).

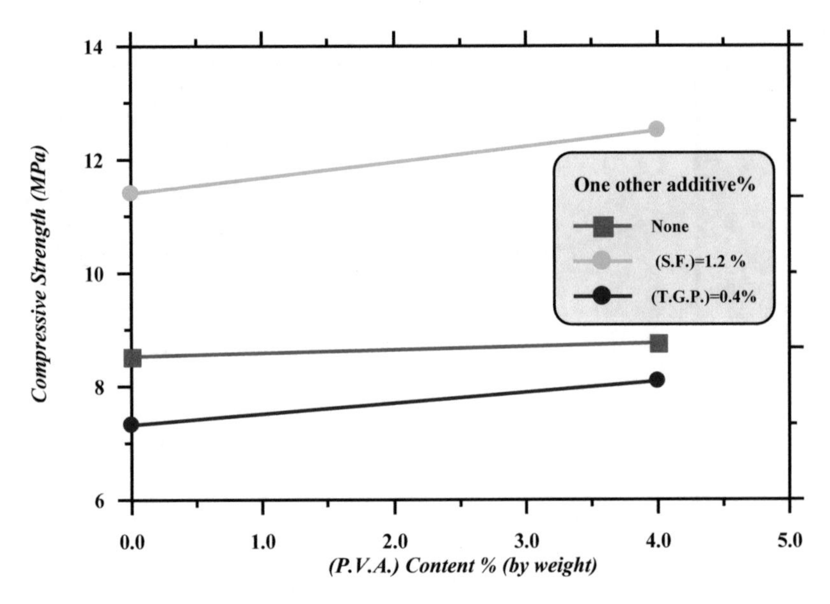

Fig. 4. Effect of (P.V.A.) on Compressive Strength of Juss with and without one other additive (S.F.) or (T.G.P.).

Table 4. Effect of (P.V.A.) on Compressive Strength of Juss with and without one other additive (S.F.) or (T.G.P.).

Mix No.	Content of one other additive by weight (%)	(P.V.A.) content by weight (%)	(W/J) ratio	Compressive strength (MPa)	Percentage of increase (%)
Mix 1	No other additive	0.0	0.3	8.52	——
Mix 3		4.0	0.3	8.75	2.69
Mix 2	(S.F.) content = 1.2%	0.0	0.3	11.4	——
Mix 5		4.0	0.3	12.5	9.64
Mix 4	(T.G.P.) content = 0.4%	0.0	0.3	7.32	
Mix 7		4.0	0.3	8.08	10.38

3.2 Setting Time

Effect of (S.F.) on Setting Time of Juss with and Without One Other Additive (T.G.P.) or (P.V.A.)

Figure 5 and Table 5 shows the effect of (S.F.) on the compressive strength of the juss. They illustrate that the compressive strength is decreased by (22.07%) when (S.F.) is added alone to the mix, while the percentage of decreasing becomes (44.39%) when (S. F.) is added together with (T.G.P.), and becomes (21.36%) when (S.F.) is added together with (P.V.A.).

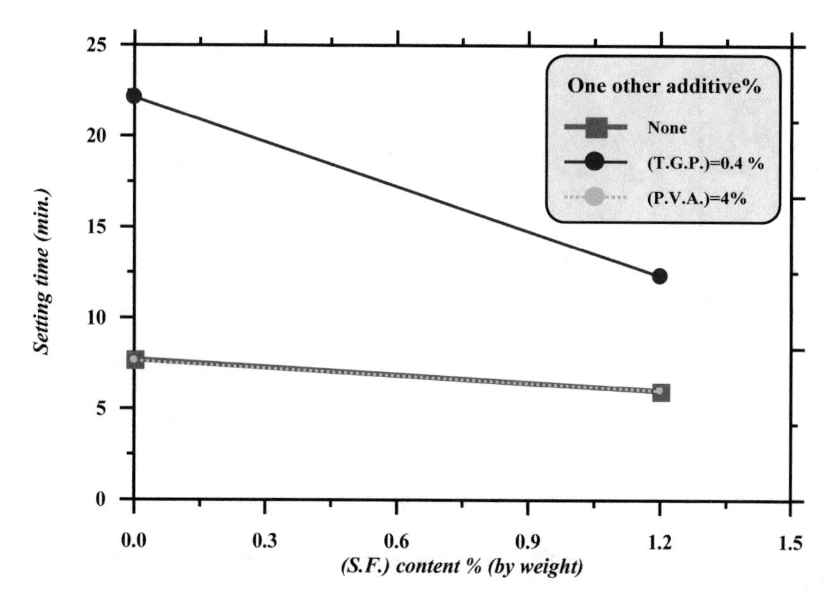

Fig. 5. Effect of (S.F.) on Setting Time of Juss with and without one other additive (T.G.P.) or (P.V.A.).

Table 5. Effect of (S.F.) on Setting Time of Juss with and without one other additive (T.G.P.) or (P.V.A.).

Mix No.	Content of one other additive by weight (%)	(S.F.) content by weight (%)	(W/J) ratio	Setting time (min.)	Percentage of decrease (%)
Mix 1	No other additive	0.0	0.3	7.7	——
Mix 2		1.2	0.3	6	22.07
Mix 4	(T.G.P.) content = 0.4%	0.0	0.3	22.12	——
Mix 6		1.2	0.3	12.3	44.39
Mix 3	(P.V.A.) Content = 4%	0.0	0.3	7.63	——
Mix 5		1.2	0.3	6	21.36

Effect of (T.G.P) on Setting Time of Juss with and Without One Other Additive (S.F.) or (P.V.A.)

Figure 6 and Table 6 shows the effect of (T.G.P.) on the compressive strength of the juss. They illustrate that the compressive strength is increased by (187.27%) when (T. G.P.) is added alone to the mix, while the percentage of increasing becomes (105.0%) when (T.G.P.) is added together with (S.F.), and becomes (188.33%) when (T.G.P.) is added together with (P.V.A.).

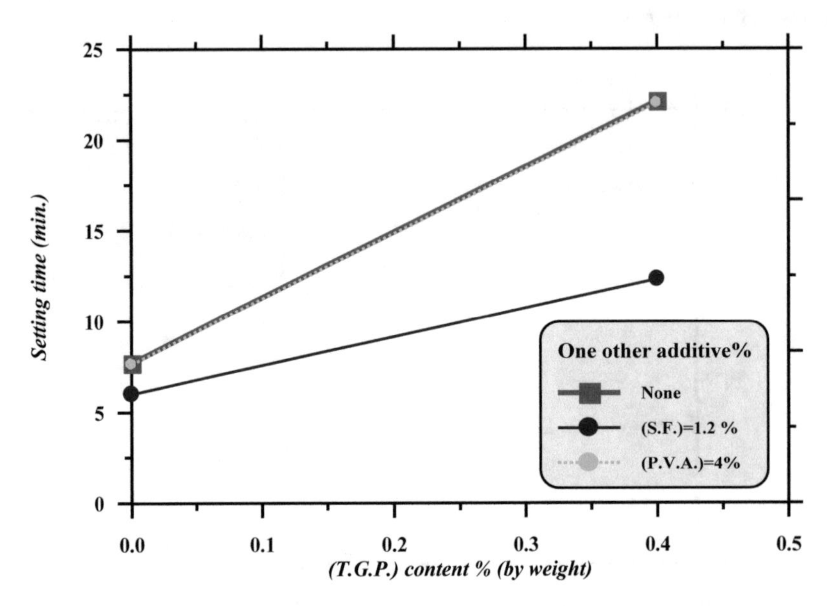

Fig. 6. Effect of (T.G.P.) on Setting Time of Juss with and without one other additive (S.F.) or (P.V.A.).

Table 6. Effect of (T.G.P.) on Setting Time of Juss with and without one other additive (S.F.) or (P.V.A.).

Mix No.	Content of one other additive by weight (%)	(T.G.P.) content by weight (%)	(W/J) ratio	Setting time (min.)	Percentage of increase (%)
Mix 1	No other additive	0.0	0.3	7.7	——
Mix 4		0.4	0.3	22.12	187.27
Mix 2	(S.F.) content = 1.2%	0.0	0.3	6	——
Mix 6		0.4	0.3	12.3	105.00
Mix 3	(P.V.A.) content = 4.0%	0.0	0.3	7.63	
Mix 7		0.4	0.3	22	188.33

Effect of (P.V.A.) on Setting Time of Juss with and Without One Other Additive (S.F.) or (T.G.P.)

Figure 7 and Table 7 shows the effect of (P.V.A.) on the compressive strength of the juss. They illustrate that the compressive strength is slightly decreased by (0.909%) when (P.V.A.) is added alone to the mix, while the percentage of decreasing becomes (0.00%) when (P.V.A.) is added together with (S.F.), and becomes (0.45%) when (P.V. A.) is added together with (T.G.P.).

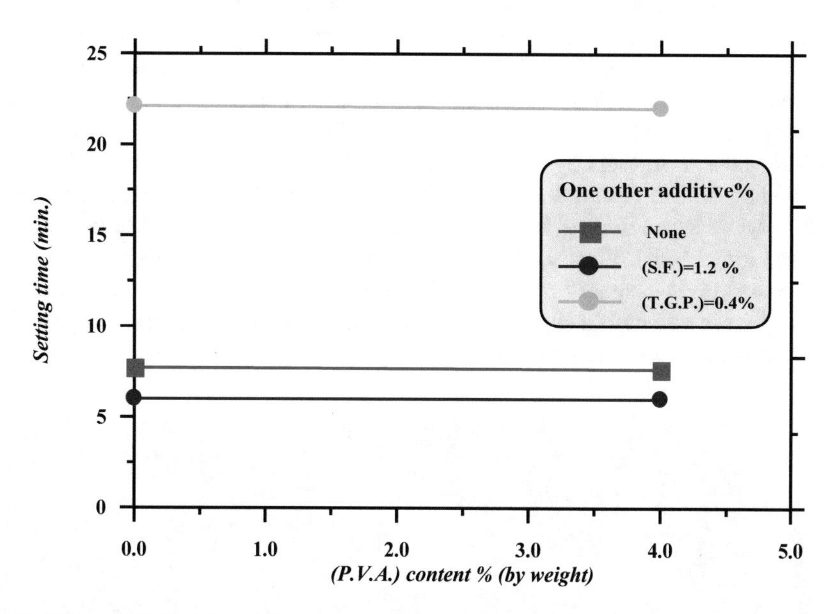

Fig. 7. Effect of (P.V.A.) on Setting Time of Juss with and without one other additive (S.F.) or (T.G.P.).

Table 7. Effect of (P.V.A.) on Setting Time of Juss with and without one other additive (S.F.) or (T.G.P.).

Mix No.	Content of one other additive by weight (%)	(P.V.A.) content by weight (%)	(W/J) ratio	Setting time (min.)	Percentage of decrease (%)
Mix 1	No other additive	0.0	0.3	7.7	——
Mix 3		4.0	0.3	7.63	0.909
Mix 2	(S.F.) content = 1.2%	0.0	0.3	6	——
Mix 5		4.0	0.3	6	0.000
Mix 4	(T.G.P.) content = 0.4%	0.0	0.3	22.12	——
Mix 7		4.0	0.3	22	0.450

3.3 Privilege of the Combined Effect of Adding: (S.F.), (P.V.A.) and (T.G.P.)

On Compressive Strength

Bar Chart 1 presents a view of the percentages of increasing or decreasing in the compressive strength of local gypsum (juss) when the additives are added as a combination of two or three, and comparing these percentages with those resulted by using each additive individually. It displays that if taking the case of Mix 1 (i.e.: zero S.F., P.V.A. and T.G.P.) as a reference case, then when adding S.F. alone (Mix 2), the compressive strength is increased by (33.8%), while when using P.V.A. alone (Mix 3), the compressive strength is increased by (2.69%), but when using T.G.P. alone (Mix 4) the compressive strength is decreased by (14.08%). On the other hand, when adding both S.F. and P.V.A. (Mix 5), the compressive strength is increased by (46.71%), and when adding both S.F. and T.G.P. (Mix 6) the compressive strength is increased by (4.22%), while when adding both (P.V.A.) and (T.G.P.) the compressive strength is decreased by (5.16%). Finally, when we add the three additives altogether, the compressive strength is increased by (30.28%). The mentioned bar chart reveals an obvious improvement in all combination results as compared with the corresponding results of the individual usage of all additives.

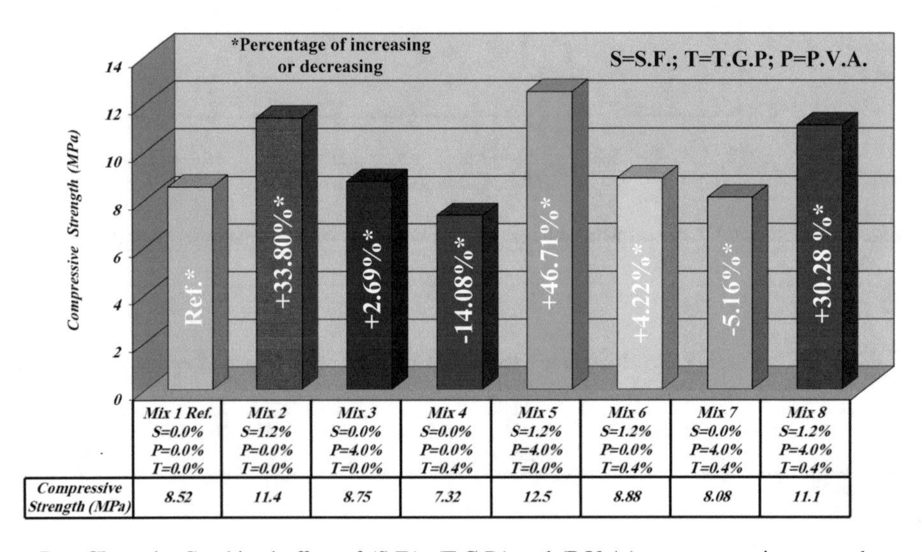

Bar Chart 1. Combined effect of (S.F.), (T.G.P.) and (P.V.A.) on compressive strength.

On Setting Time

Bar Chart 2 presents a view of the percentages of increasing or decreasing in the setting time of local gypsum (juss) when the additives are added as a combination of two or three, and comparing these percentages with those resulted by using each additive individually. It displays that if taking the case of Mix 1 (i.e.: zero S.F., P.V.A. and T.G.P.) as a reference case, then when adding S.F. alone (Mix 2), the compressive strength is decreased by (22.07%), and when using P.V.A. alone (Mix 3), the compressive strength is decreased by (0.909%), but when using T.G.P. alone (Mix 4) the compressive strength is increased by (187.27%). On the other hand, when adding both S.F. and P.V.A. (Mix 5), the compressive strength is decreased by (22.07%), while when adding both S.F. and T. G.P. (Mix 6) the compressive strength is increased by (59.74%), and when adding both (P.V.A.) and (T.G.P.) the compressive strength is increased by (185.71%). Finally, when we add the three additives altogether, the compressive strength is increased by (75.32%). The mentioned bar chart reveals an improvement in almost all combination results as compared with the corresponding results of the individual usage of all additives.

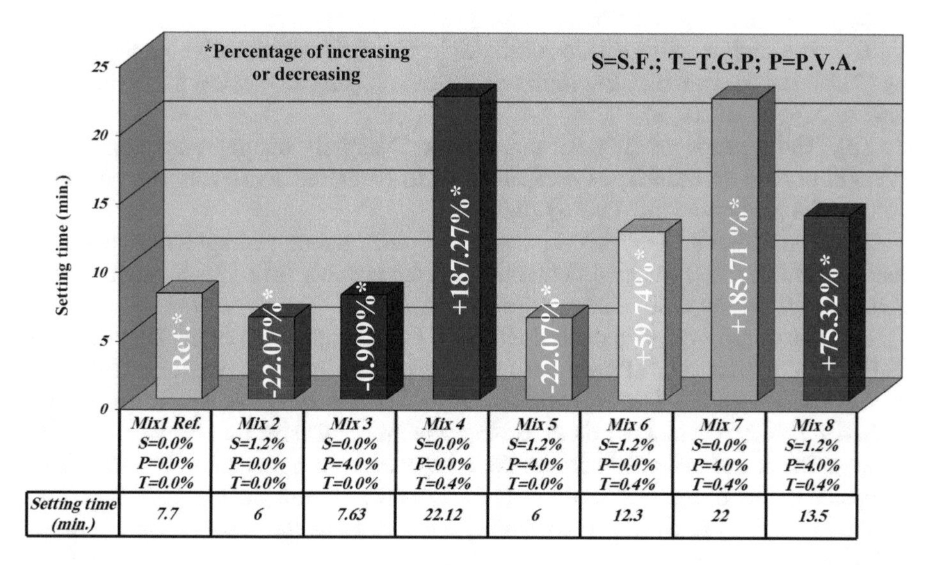

Bar Chart 2. Combined effect of (S.F.), (T.G.P.) and (P.V.A.) on setting time.

4 Conclusions

(1): (A): From Table 2 we can conclude that when adding (S.F.) alone, the compressive strength is improved by (33.8%), and this percentage of improvement is reduced with the presence of (T.G.P.), and enlarged with the presence of (P.V.A.).

(B): From Table 5 we conclude that when adding (S.F.) alone, the setting time is reduced by (22.07%), and this percentage of reduction is increased with the presence of (T.G.P.), and decreased with the presence of (P.V.A.).

(2): (A): From Table 3 one can conclude that when adding (T.G.P.) alone, the compressive strength is reduced by (14.08%), and this percentage of reduction is increased with the presence of (S.F.), and decreased with the presence of (P.V.A.).

(B): From Table 6 we can conclude that when adding (T.G.P.) alone, the compressive strength is highly improved by (187.27%), and this percentage of improvement is reduced with the presence of (S.F.), and remains almost constant with the presence of (P.V.A.).

(3): (A): From Table 4 we may conclude that when using (P.V.A.) alone, the compressive strength of (Juss) is slightly increased by (2.69%), and this percentage of increasing is enlarged with the presence of (S.F.), and enlarged a little more with the presence of (T.G.P.)

(B): From Table 7 we might conclude that when using (P.V.A.) alone, the setting time of (Juss) is slightly reduced by (0.909%), and (P.V.A.) as no effect on setting time when used with (S.F.), and the setting time is decreased a little more by (0.45%) when (P.V.A) is used with (T.G.P.).

(4): When considering the effect of the individual use of the three additives (S.F.), (P.V.A.) and (T.G.P.) on the two studied properties of local gypsum (Juss) simultaneously from noticing Bar Charts 1 and 2, we notice the following outcomes:

(A): The addition of (S.F.) alone (Mix 2) improves the compressive strength of Juss by (33.8%) as compared to that of the reference mix (Mix 1), but it reduces the setting time of Juss by (22.07%).

(B): The addition of (P.V.A.) alone (Mix 3) slightly improves the compressive strength of Juss by (2.69%) as compared to that of the reference mix (Mix 1), but it reduces the setting time of Juss by (0.909%).

(C): The addition of (T.G.P.) alone (Mix 4) decreases the compressive strength of Juss by (14.08%) as compared to that of the reference mix (Mix 1), although it highly increases the setting time of Juss by (187.27%).

(5): In order to minimize the bad effects of the individual use of the three additives (S.F.), (P.V.A.) and (T.G.P.),we went to the dual use of these additives (Mix 5: S. F. + P.V.A = Mix 2 + Mix 3), (Mix 6: S.F. + T.G.P. = Mix 2 + Mix 4) and (Mix 7: P.V.A. + T.G.P. = Mix 3 + Mix 4). From studying Bar Charts 1 and 2, we notice that:

(A): The use of (Mix 5) improves the compressive strength of Juss by (46.71%) as compared to that of the reference mix (Mix 1) which exceeds the algebraic sum of the results of (Mix 2) & (Mix 3)individually, but it reduces the setting time of Juss by (22.07%).

(B): The use of (Mix 6) slightly improves the compressive strength of Juss by (4.22%) as compared to that of the reference mix (Mix 1), and it also increases the setting time of Juss by (59.74%).

(C): The use of (Mix 7) slightly reduces the compressive strength of Juss by (5.16%) as compared to that of the reference mix (Mix 1), but it highly increases the setting time of Juss by (185.71%).

6): Finally, and for the aim of gaining the maximum benefit of the three additives and minimizing their bad effects (simultaneously), we went to the triple use of these additives (Mix 8: S.F. + P.V.A. + T.G.P. = Mix 2 + Mix 3 + Mix 4). From Bar Charts 1 and 2, we notice that: The use of (Mix 8) improves the compressive strength of Juss by (30.28%) as compared to that of the reference mix (Mix 1) which exceeds the algebraic sum of the results of (Mix 2), (Mix 3) & (Mix 4) individually, and it also improves the setting time of Juss by (22.07%). This simultaneous improvements in the two studied properties of Juss (namely; the compressive strength and the setting time) provided by (Mix 8) which is the mix of all three additives presents the best outcomes gained as compared with all previous mixes outcomes.

References

1. Khalil, A.A., Gad, G.M.: Mineral and chemical constitutions of the UAR. Indian Ceramics **16**, 173–177 (1972). Cited by reference [9]
2. Combe, E.C., Smith, D.C.: Some properties of gypsum plaster. J. Brit. Dent. **17**, 237–245 (1964). Cited by reference [9]
3. Peters, C.P., Hines, J.L., Bachus, K.N., Craig, M.A., Bloebaum, R.D.: Biological effect of calcium sulfate as bone graft substitute in ovine metaphyseal defects. J. Biomed. Mater. Res. A. **76**(3), 456–462 (2005). Cited by reference [9]
4. Craig, R.G.: Restorative Dental Materials, 7th edn., pp. 303–330. The C.V. Mospy comp., St., Louis (1989). Cited by reference [9]
5. Papageorgiou, A., Tzouvalas, G., Tsimas, S.: Use of inorganic setting retarders in cement industry. Cem. Concr. Res. **27**, 183–189 (2005). Cited by reference [9]
6. El-Maghraby, H.F., Gedeon, O., Khalil, A.A.: Formation and Characterization of Poly(vinyl alcohol – co – vinyl Acetate – co-itaconic Acid)/Plaster Composites: part II: composite formation and characteristics. Ceramic Silikaty **51**(3), 168–172 (2007). Cited by reference [9]
7. Bas_pinar, S.M., Kahraman, E.: Modifications in the properties of gypsum construction element via addition of expanded macroporous silica granules. Constr. Build. Mater. **25**, 3327–3333 (2011). Cited by reference [9]. http://dx.dioi.org/10.10016/j.coinbuildmat.2011.03.022
8. Khalil, A.A., Abdel-kader, A.H.: Preparation and physicomechanical properties of gypsum plaster-agro fiber wastes composites. Interceram Int. J. Refract. Manual (Spec. Technol.) **21**, 62–67 (2010). Cited by reference [9]
9. Khalil, A.A., Tawfik, A., Hegazy, A.A., El-Shahat, M.F.: Effect of different forms of silica on the physical and mechanical properties of gypsum plaster composites. Materiales deConstrucción **63**, 312, 529–537 (2013)
10. Murat, M., Attari, A.: Modification of some physical properties of gypsum plaster by addition of clay minerals. Cem. Concr. Res. **2**, 378–387 (1991). Cited by reference [9]
11. Wen, L., Yu-he, D., Mei, Z., Ling, X., Qian, F.: Mechanical properties of nano SiO2 filled gypsum particle board. Trans. Nonferrous Metal. Soc. China **16**, 361–364 (2006). Cited by reference [9]
12. Fu, X., Chung, D.D.L.: Effects of silica fume, latex, methylcellulose, and carbon fibers on the thermal conductivity and specific heat of cement paste. Cem. Concr. Res. **27**(12), 1799–1804 (1997). Cited by reference [9]
13. Shebl, S.S., Seddeq, H.S., Aglan, H.A.: Effect of micro-silica loading on the mechanical and acoustic properties of cement pastes. Const. Build. Mater. **25**, 3903–3908 (2011). Cited by reference [9]. http://dx.dioi.org/10.1016/j.conbiuildmat.2011.04.021
14. Ogawa, K., Uchikiawa, H., Takemoto, K., Yasui, I.: The mechanism of the hydration in the system C3S-pozzolana. Cem. Concr. Res. **10**, 683–696 (1980). Cited by reference [9]
15. Wirsching, F.: Drying and agglomeration of flue gas gypsum. In: Kuntze, R.A. (ed.) The Chemistry and Technology of Gypsums Philadelphian, American Society for Testing and Materials, pp. 161–174 (1984). Cited by reference. Padevět, P. Tesárek, T. Plachý "Evolution of mechanical properties of gypsum in time", INTERNATIONAL JOURNAL OF MECHANICS, Issue 1, Volume 5, 2011
16. Aljubouri Auday, Z.A., Othman, O: Physical properties and compressive strength of the technical plaster and local Juss. Iraqi J. Earth Sci. **9**(2), 49–58 (2009)
17. ASTM C472 – 99 (Reapproved 2009): Standard Test Methods for Physical Testing of Gypsum Plaster and Gypsum Concrete. Annual Book of ASTM Standard, September 1 2009

Evaluation of Tensile Strength and Durability of Microbial Cement Mortar

Ahmed S. D. AL-Ridha[1(✉)], Ali F. Atshan[2], Hussein H. Hussein[3], Ali A. Abbood[1], Layth Sahib Dheyab[4], and Ayoob Murtadha Alshaikh Faqri[5]

[1] Structural Engineering, Department of Civil Engineering, College of Engineering, Mustansiriyah University, Baghdad, Iraq
ahmedsahibdiab@yahoo.com

[2] Structural Engineering, Department of Water Resources Engineering, College of Engineering, Mustansiriyah University, Baghdad, Iraq

[3] Departments Petroleum Engineering, College of Engineering, Baghdad University, Baghdad, Iraq

[4] Civil Engineering, Consultant Engineer, Baghdad, Iraq

[5] Biotechnology, Baghdad University, Baghdad, Iraq

Abstract. In this research, an attempt has been made to study the effect of adding the micro-organisms in cement mortar on its tensile strength. In addition to evaluating the durability of cement mortar with and without micro-organisms by studying the deterioration in tensile strength after sulfuric acid (H_2SO_4 with 6% concentration) attack.

The experimental work included casting four mixes (denoted by: case 1, case 2, case 3, case 4 and the control mixes) which differ according to the way of adding micro-organisms, casting water, curing water (or both).

Each mix contains three cylindrical specimens except case 4 and the control mix which contain six cylinders each. It was found that, when adding micro-organisms; the tensile strength was increased (i.e. all mixes gave a higher percentage of tensile strength) as compared with that of the control mix.

The results revealed that for the microbial cement mortar (case 4), the percentage of deterioration in tensile strength, was smaller [after the sulfuric acid H_2SO4 (6% concentration) attacked the samples] than that of the control mix. It was also found that the effect of using bacterial additives on the percentage of increasing in tensile strength of cement mortar (case4 & control) is increased after sulfuric acid (H_2SO_4: 6% concentration) attack.

Keywords: Micro-organisms · Bio-mineralization · Microbial cement mortar · Tensile strength and durability

1 Introduction

Available literature shows that micro-organisms' materials have an essential part to improve mechanical properties of cement mortar. They precipitate and fill the pores of the mix, which increase the strength and durability [1–5]. In addition, in specific cases, microorganisms' materials managed to make a transformation for the periodic table elements [5].

M. Shehata et al. (Eds.): GeoMEast 2019, SUCI, pp. 80–89, 2020.
https://doi.org/10.1007/978-3-030-34249-4_8

These findings lead to a new term that called "bio-mineralization process". The applications of this process expected to be used widely in construction and in other new materials [6–8].

Moreover, many studies have shown that bacterial calcium carbonate precipitation can used to improve the tensile strength of mortar [6, 7, 9–12]. They encouraged using the microbial materials that produce the calcite precipitation as a green material, these actives are natural and pollution-free.

Durability is the ability of material to last a long time without deterioration. In general, durability of cement mortar depends on permeability of mix [13]. The permeability depends on the porosity and on the pores structure. The addition open the pore of materials high penetrability was a reason of degradation of materials. The open pore allows to high possibility of penetrating by aggressive substances [14]. Therefore, mechanical properties, such as the test of tensile strength may be required to evaluate concrete durability [15]. There is a need for further research focuses on understanding relation between adding organic material and the compression strength and durability of material.

2 Experimental Work

2.1 Materials

Cement
Ordinary Portland local cement (Type–I) was used in all mixes throughout this research.

Fine Aggregate
The characteristics of fine aggregate (sand) were; (4.75 mm) maximum size, (2.6) specific gravity, smooth texture and rounded shape particles. The fine aggregate were collected locally from Al - Ukhaidhir field.

Mixing water
Normal drinking water was used for mixing and curing in all samples of this study.

Limestone Powder (LSP)
It is a white grinded material of limestone collected locally in Iraq. Grinding of this material is carried out by blowing technique until it becomes a fine powder.

Silica Fume (S.F.)
Silica fume is highly reactive pozzolanic substance and is a byproduct from the production of silicon or ferro- silicon metal. It is a very fine powder and composed from the flue gases from electric arc furnaces. The silica fume used in is imported from Egypt.

Eggshell
Eggshell powder is used in all cement mortar mixes.

Micro-organisms
The Bacteria used in this study was: *Bacillus subtilis* ATCC 6633. Table 1 shows the Microscopic and Phenotypic features of *Bacillus subtilis:*

Table 1. Microscopic and phenotypic features of *Bacillus subtilis*

Microscopic Features	Gram-Postive bacilli
Phenotypic Features	Results
Raffinose	–
Rhamnose	–
Salicin	+
Sorbitol	+
Sucrose	+
ONPG	+
Nitrate	+
Arginine Dihydrolase	–
Citrate Utilisation	–
Other Features	Results
H_2O_2	+

Note: "+":- Present "-":- Absent

- Culture of *Bacillus subtilis*

The pure culture was isolated from the soil sample of HKM. It had been cultivated in a laboratory using a nutrient agar slants. After 24 h, it forms colonies in nutrient agar slants which were maintained in a test tube. Then the test tube is kept at a refrigerator providing (4 °C) temperature, till further use.

- Maintenance of Stock Cultures

Whenever it is required to use micro-organisms with a single colony of the culture, take swab from test tube which has (nutrient agar slants + *Bacillus subtilis*) by using a sterile loop and put in a glass bottle which had medium broth of 13 g/L (This research take the initiative to test the behavior of bacteria on the reaction without adding reagent material such as urea). These glass bottles were kept at warm temperature 37 °C for 48 h in laboratory incubation. When growth becomes noticeable, it will be compared with the MacFarland tube 0.5. This tube prepared to determine the concentration of bacteria per ml. When the cell concentrations become 1×10^7 cells/ml, the glass bottles preserved at refrigerator temperatures (under 4 °C) until usage in the present work. Figure 1 shows the glass bottles that contains the *Bacillus subtilis.*

Fig. 1. Glass bottles that contains the *Bacillus subtilis*

2.2 Cement Mortar Mixes

The mix proportions (by weight) of cement mortar; (cement: sand: limestone powder) was; (1:1. 9:0.1) respectively. The water-cement ratio was chosen to be (0.5). The contents of Silica fume and eggshell powders added to the mixes were (3%) and (1%) by weight of cement, respectively.

Ways of Adding Micro-organisms to Mixes
In this experimental study, micro-organisms were added to the four cases of cement mortar, by three ways:-

1- For [case 1, case 2] mixes: By adding them to mixing water which will have a cell concentration 1×10^7cells/ml and Nutrient broth (13 g/L).

2- For [case 3] mix: After casting, where Micro-organisms are mixed with the curing water which will have Nutrient broth (13 g/L) + microorganisms and then sprinkled to the hardened mortar.

3- For [case 4] mix: Using both ways that were explained in points (1) and (2) together Table 2 illustrates the above mentioned process.

Table 2. Description of mix cases.

No. of case	Cell Concentration/ml of mixing water	Types of curing waters
Control	Nil (control)	Normal water with no additive
Case 1	1×10^7	Normal water with no additive
Case 2	1×10^7	Water + Nutrient broth
Case 3	Nil	Water + Nutrient broth + microorganisms[a]
Case 4	1×10^7	Water + Nutrient broth + microorganisms[a]

[a]unknown cell concentration

Mixing Procedure
In the beginning, dry materials such as; cement, sand, limestone, silica fume and eggshell powders are mixed in a pan for several minutes. After that, the needed amount of water is added to the mix with the required content of micro-organisms according to the process explained in the previous article with the four cases of cement mortar. Mixing continues until a homogeneous fresh cement mortar is obtained.

Casting and Curing
When the mixing process was finished, the fresh cement mortar was then poured into the cylinders mold (shown in Fig. 2) and vibrated by putting it over an electric generator machine which provides the required vibration. Twenty–four hours later, samples were stripped out of the molds. Then, samples were cured using three types of water for the four cases (besides the control mix) to achieve the curing for the cement mortar in the present work.These types of water are:-

Fig. 2. Specimens' Molds; Cement Mortar Specimens (With Micro-organisms; Without Micro-organisms; Inside Acid Attack; After Acid Attack); Testing Machine and Some tested specimens.

1- A curing water without any additive [used for the control mix and case 1].

2- A curing water which have Nutrient broth (13 g/L) [used for case 2].

3- A curing water which have (13 g/L) Nutrient broth + microorganisms [used for case 3 and case 4].

These types of curing waters are shown in Table 2.

The curing was at laboratory conditions, temperature of water was between 25 °C and 30 °C. Samples kept in curing water for a month. Then, they taken out from the curing water and were ready to test as seen in the following paragraphs.

2.3 Tensile Strength Test

The specimens used in this research are all cylindrical specimens (50 × 100) mm dimensions, as shown in Fig. 2, and they are brought to be tested when they attain the age of about one month. The test was carried out using electrical testing machine with capacity of (2000 kN), and the tensile strength test was achieved according to ASTM C496-86 [16].

3 Results and Discussion

3.1 Effect of Additive Micro-organisms on Tensile Strength of Cement Mortar with Variable Mix Cases

In this work, three types of added Bacteria *Bacillus subtilis* in cement mortar are studied with three types of curing waters ([case 1], [case 2], [case 3], [case 4] and control mixes) as illustrated in Table 2.

Table 3 and Fig. 3 show the effect of the "Bacteria *Bacillus subtilis*" additive on the tensile strength of cement mortar. The results reveal that [case 4] mix gives the higher percentage of increase.

Table 3. Effect of additive micro-organisms on tensile strength of cement mortar with variable mix case

Mix designation	Tensile strength MPa	Percentage of increase%
Control	2.77	–
Case 1	3.0	8.30
Case 2	3.11	12.27
Case 3	3.02	9.03
Case 4	3.25	17.33

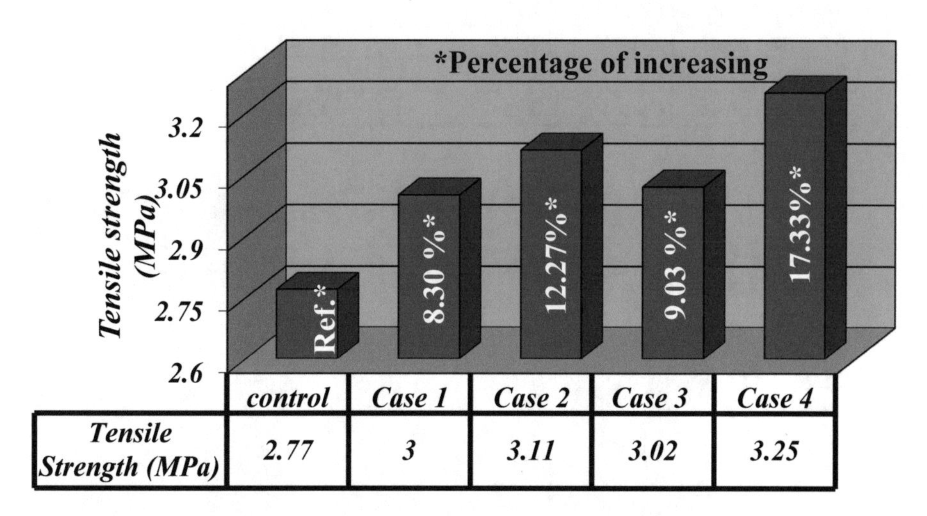

Fig. 3. Effect of additive microorganisms on tensile strength of cement mortar with variable cases

This improvement in the tensile strength of cement–sand mixture depends on the structural shape of the pores, whether $CaCO_3$ deposits in or fills these pores. As a result of the bacterial growth, the calcium deposition process occurs continuously, clogging the internal pores with calcium deposits [17].

3.2 Durability

The Experimental work was focusing mainly on the effect of Sulfuric acid attack on the tensile strength of cement mortar specimens with and without Bacteria [control & case 4 mixes]. Specimens were immersed in 6% solutions of H_2SO_4. The specimens are arranged in the plastic tubs for about one week. Before testing, each specimen was removed from the bath, and rinsed under water tap.

Effect of Sulfuric Acid on Tensile Strength of Cement Mortar with and without Microorganisms

Table 4 and Fig. 4 show the effect of the sulfuric acid (H_2SO_4: 6% concentration) attack on the deterioration of the tensile strength of cement mortar with and without Micro-organisms (case 4 & control mixes). The results illustrate, that the deterioration of tensile strength of cement mortar due to sulfuric acid (H_2SO_4) attack has decreased with the presence of micro-organisms. This behavior may be imputed to deposition of CaCO3 generated by micro-organisms in the voids or pores within cement mortar specimens, which lead to increase the durability and hence increasing the resistance to sulfuric acid (H_2SO_4) attack.

Table 4. Effect of sulfuric acid on tensile strength of cement mortar with and without microorganisms

Test condition	No. of case	Tensile strength MPa	Percentage of decrease %
Before acid attack	Control	2.77	–
After acid attack	Control	2.23	19.49
Before acid attack	Case 4	3.25	–
After acid attack	Case 4	2.83	12.92

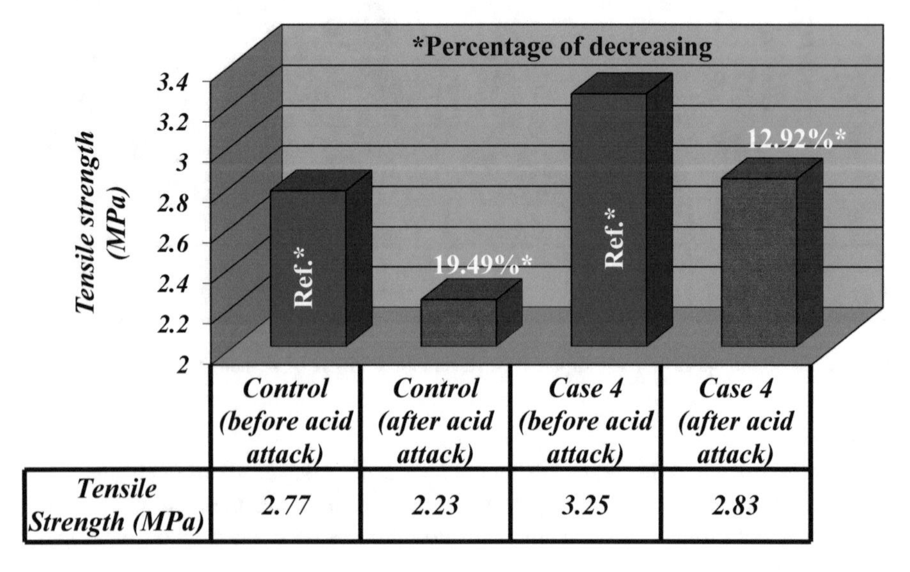

Fig. 4. Effect of sulfuric acid on tensile strength of cement mortar with and without microorganisms

Effect of Micro-organisms Additive on Tensile Strength of Cement Mortar with and without Sulfuric Acid

Table 5 and Fig. 5 show the effect of using bacterial additive on tensile strength of cement mortar with and without attack sulfuric acid (H_2SO_4) (6%) (case 4 & control mixes). The results reveal that, the tensile strength of cement mortar increases in the presence of microorganisms, and the percentage of this increase is magnified with sulfuric acid (H_2SO_4: 6% concentration) attack. This behavior may be attributed to the deposition of $CaCO_3$ by micro-organisms in the voids or pores existed within cement mortar mass, which lead to increase the durability and accordingly increase the resistance to sulfuric acid (H_2SO_4) attack.

Table 5. Effect of additive microorganisms on tensile strength of cement mortar with and without sulfuric acid

Test condition	No. of case	Tensile strength MPa	Percentage of increase %
Before acid attack	Control	2.77	–
	Case 4	3.25	17.32
After acid attack	Control	2.23	–
	Case 4	2.83	26.90

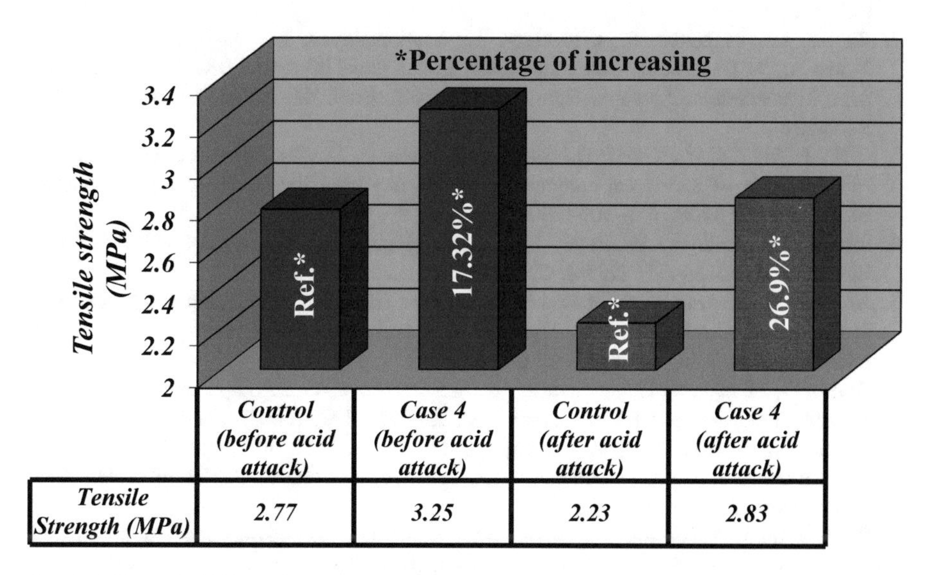

Fig. 5. Effect of additive microorganisms on tensile strength of cement mortar with and without sulfuric acid

4 Conclusion

1- When adding micro-organisms to the cement mortar in different ways, through four cases (casting(mixing) water, curing water or both), the tensile strength is improved. And Case (4) mix gives the higher parentage of increasing.

2- When immersing cement mortar in sulfuric acid (H_2SO_4: 6% concentration) for a week period (as an indicator to the durability of cement mortar) in order to evaluate the degradation of tensile strength, it was found that:-

   a- The Effect of Sulfuric acid on degradation of tensile strength of cement mortar is decreased in microbial cement mortar (case4 mix) compared with the conventional cement mortar.

   b- The effect of using bacterial additives on the percentage of increasing in tensile strength of cement mortar (case4 & control) is increased after sulfuric acid (H_2SO_4: 6% concentration) attack.

References

1. Bang, S.S., Galinat, J.K., Ramakrishnan, V.: Calcite precipitation induced by polyurethane-immobilzed Bacillus pasteurii. Enzyme Microb. Technol. **28**, 404–409 (2001). Cited by Reference 5
2. Biswas, M., Majundar, S., Chowdury, T., Chattopadhyay, B., Mandal, S., Halder, U., Yamasaki, S.: Bioremediase aunique protein from a novel bacterium BKH1, ushering a new hope in concrete technology. Enzyme Microb. Technol. **46**, 581–587 (2010). Cited by Reference 5
3. Dick, J., Windt, W., Graef, B., Saveyn, H., Meeren, P., De Belie, N., Verstraete, W.: Biodeposition of a calcium carbonate layer on degraded limestone by bacillus species. Biodegradation **17**(4), 357–367 (2006). Cited by 5
4. Fischer, S.S., Galinat, J.K., Bang, S.S.: Microbiological precipitation of CaCO3. Soil Biol. Biochem. **31**, 1563–1571 (1999). Cited by Reference 5
5. Afifudin, H., Nadzarah, W., Hamidah, M.S., Noor Hana, H.: Microbial participation in the formation of Calcium Silicate Hydrated (CSH) from *Bacillus subtilis*. In: The 2nd International Building Control Conference (2011)
6. Ghosh, P., Mandal, S., Chattopadhyay, B.D., Pal, S.: Use of microorganism to improve the strength of cement mortar. Cem. Concr. Res. **35**, 1980–1983 (2005). Cited by Reference 11
7. Ghosh, S., Biswas, M., Chattopadhya, B.D., Mandal, S.: Microbial activity on the microstructure of bacteria modified mortar. Cem. Concr. Compos. **31**, 93–98 (2009). Cited by Reference 11
8. Schultze-Lam, S., Fortin, D., Davis, B.S., Beveridge, T.J.: Mineralization of bacterial surfaces. Chem. Geol. **132**, 171–181 (1996). Cited by Reference 11
9. Achal, V., Mukherjee, A., Basu, P.C., Reddy, M.S.: Strain improvement of Sporosarcina pasteurii for enhanced urease and calcite production. J. Ind. Microbiol. Biotechnol. **36**, 981–988 (2009). Cited by Reference 11
10. Bang, S.S., Galinat, J.K., Ramakrishnan, V.: Calcite precipitation induced by polyurethane-immobilized Bacillus pasteurii. Enzyme Microb. Technol. **28**, 404–409 (2001). Cited by Reference 11

11. Ramachandran, S.K., Ramakrishnan, V., Bang, S.S.: Remediation of concrete using micro-organisms. ACI Mater. J. **98**, 3–9 (2001)
12. Park, S.-J., Park, Y.-M., Chun, W.-Y., Kim, W.-J., Ghim, S.-Y.: Calcite-forming bacteria for compressive strength improvement in mortar. J. Microbiol. Biotechnol. **20**(4), 782–788 (2010)
13. Khan, M.I.: Isoresponses for strength, permeability and porosity of high performance Mortar. Build. Environ. **38**, 1051–1056 (2003). Cited by Reference 15
14. Claisse, P.A., Elsayad, H.A., Shaaban, I.G.: Absoprtion and sorptivity of cover concrete. J. Mater. Civ. Eng. **9**, 105–110 (1997). Cited by Reference 15
15. Achal, V., Mukherjee, A., Sudhakara Reddy, M.: Microbial concrete: a way to enhance durability of building structures. In: Second International Conference on Sustainable Construction Materials and Technologies June 28 - June 30, 2010, Università Politecnica delle Marche, Ancona, Italy (2010)
16. ASTM C 496–86: Standard test method for splitting tensil of cylindrical concrete specimens. Annual Book of ASTM Standards, **04**(02), 259–262 (1989)
17. Seshagiri Rao, M.V., Srinivasa Reddy, V., Sasikala, Ch.: Performance of microbial concrete developed using bacillus subtilus JC3. J. Inst. Eng. India Ser. A (2017)

Bio Concrete

Hager El-Mahdy(✉) and Ahmed Tahwia

Faculty of Engineering, Civil Engineering Department, Mansoura University, Mansoura, Egypt
Hageralmahdy8@gmail.com, atahwia@yahoo.com

Abstract. Concrete is the most widely used construction material works in the world, but it dose have a weakness it is prone to catastrophic cracking that has immense financial and environmental impact. Bio concrete is an environment friendly and innovative approach aiming to close the cracks produced in concrete structures. This paper presents the engineering characteristics of two types of bacteria; *Sporosarcina pasteurii*, and *Rhizobium leguminosarum*, mixed with concrete with different concentrations. The experimental laboratory testing included compressive strength, split tensile strength, flexural strength, and ultrasonic pulse velocity test, CT-Scanning, SEM test, EDX test, to investigate the bacteria's effect on the strength of concrete at the ages of 7, and 28 days. In general, it was found that the addition of bacteria proved to have higher strength characteristics by closing the micro pores in concrete and making it denser as compare to the conventional concrete.

Keywords: Bacterial concrete · *Sporosarcina pasteurii* · *Rhizobium leguminosarum* · Strength

1 Introduction

Concrete is a combined building material, consists of cement, coarse and fine aggregate. Cement plays an important role in the production of mortar and concrete due to its binding properties, and thus acts as a major constructional material of choice in building and structures. Rapid industrialization and urbanization increases the demand of building and construction material for infrastructure development, thereby, continuously driving the cement industry to keep growing. The increasing cement generation is also associated with certain challenges such as energy and resource conservation, cost of production, green house gas emissions, etc. It is estimated that production of Portland cement clinker alone contributes 7% global CO2 emissions, therefore, concrete does not appear to be a sustainable material [1]. The most drawback of these materials is that, it is weak in tension and so it cracks under sustained loading. Concrete cracks can be there at areas which are difficult to access and since manual repairs are exorbitant, precipitation of CaCO3 induced by bacteria. Due to the metabolism reaction, urease is produced by bacteria, whereas urease again catalyses urea and produces CO2 and NH3, and increase of the pH and carbonate concentration in the bacterial environment, resulting in ions of Calcium and Carbonate that precipitate as CaCO3 which seals the cracks.

M. Shehata et al. (Eds.): GeoMEast 2019, SUCI, pp. 90–100, 2020.
https://doi.org/10.1007/978-3-030-34249-4_9

$CO(NH_2)_2 + H_2O \rightarrow NH_3 + H_2CO_3$
$2NH_3 + 2H_2O \leftrightarrow 2NH_4^+ + 2OH^-$
$2OH^- + H_2CO_3 \leftrightarrow CO_3^{-2} + 2H_2O$
$CO_3^{-2} + Ca^{+2} \rightarrow 2NH^{+4} + CaCO_3^{-2}$

These cracks provide a pathway for harmful substances such as chlorides, carbon dioxide, sulphate, freeze and thaw cycles and ultimately oxygen and water to get in to the reinforcement, cause corrosion, rust resulting in deterioration of concrete. The use of bacterial cells has improved the strength and durability of concrete, closed the micro pres in concrete and making it denser. Literature review studies show almost all bacteria are capable of precipitating calcium carbonate, to have higher benefits the selected bacteria need to be alkaliphilic [2]. The current research revolved around improving the concrete strength by the two bacterial strains: *Sporosarcina pasteurii* and *Rhizobium leguminosarum*.

Sporosarcina pasteurii and *Rhizobium leguminosarum* are known for their diverse biochemical activities including urea hydrolysis and calcium carbonate precipitation (bio-cementing) and production of highly viscous slimy layer which causes bio-clogging, respectively.

2 Research Objectives

The main objectives of this research work is to evaluate characteristics of the bacterial concrete and determine the optimum percentage of bacteria required for bacterial concrete.

3 Materials

(1) Cement

Ordinary Portland Cement type (CEM I 52.5 N) was used. Testing conforming to of cement was carried out according to Egyptian Standard (ES 4756-1, 2013) [3]. Specific gravity of used cement was 3.15.

(2) Fine Aggregate

Natural sand with 4.75 mm maximum size composed of siliceous materials was used as Fine Aggregate (FA) in this study. Testing of sand was carried out according to the (ES 1109, 2008). Specific gravity of used sand was 2.57 and bulk density 1.76 t/m^3.

(3) Course Aggregate

Coarse Aggregate (CA) was used as crushed dolomite with a maximum nominal size of 19.5 mm in this study. The aggregate used was free from dust particles, vegetation, organic matters, and clay. Testing of gravel Aggregate was carried out according to the (ES 1109, 2008). Specific gravity of used gravel was 2.63 and bulk density 1.57 t/m^3.

(4) **Water**

Natural portable water considered as the important ingredient of concrete used for mixing and curing work. It should be clean and free from harmful impurities such as oil, acid, alkali etc.

(5) **Bacteria**

Two bacterial strains, *Sporosarcina pasteurii* (Egyptian Microbial Culture Collection (EMCC) 1960) (bio-cemented producing bacteria) and *Rhizobium leguminosarum* (EMCC 1130) (biofilm producer), were used.

4 Methodology

(1) **Procuring of Bacteria**

The bacteria were procured from Microbiological Resources Centre (Cairo MIRCEN), Ain Shams University, Cairo, Egypt. The growth and viability of the bacteria were checked in the Microbiological Laboratory, Faculty of Pharmacy, Mansoura University.

(2) **Mixing**

Cement, fine aggregate and coarse aggregate were mixed in the dry state then the water and bacteria were added. The bacteria were blended with concrete mix with percentages of 0%, 5%, 10%, and 20% by total mass of the water.

(3) **Casting**

Four blends of concrete cubes, cylinders and prisms were cast with and without bacteria. Each blend consists of 6 cubes, 3 cylinders and 3 prisms. The 4 different blends of bacterial concrete specimens were 0%, 5%, 10%, and 20% by total mass of the water. The total number of cubes cast, cylinders and prisms were 48, 24 and 24 respectively.

(4) **Curing**

Potable water was used for 7 and 28 days of curing.

(5) **Testing**

Compressive strength test on concrete cubes (15 * 15 * 15 cm) was done on 7th and 28th day for each mix. Split tensile strength test on concrete cylinders (15 * 30 cm) was conducted on 28th day. Flexural strength test on concrete prisms (10 * 10 * 50 cm) was conducted on 28th day. Also the Ultrasonic Pulse Velocity test was conducted on concrete cubes on 28th day. All tests were conducted according to ECP 203-2017 [12] at Laboratory of properties of materials, Faculty of Engineering, Mansoura University. CT-Scanning test on concrete cubes (15 * 15 * 15 cm) was done on 28th day. Scanning electron microscope (SEM) and (EDX) test on concrete cubes (15 * 15 * 15 cm) was done on 65th day.

5 Results and Discussion

(1) Compressive Strength

The compressive strength test results of all concrete mixes are presented in Table 1. The results indicate that, when Sporosarcina pasteurii and Rhizobium leguminosarum are used for partial replacement of water in concrete, have higher strength than conventional concrete as shown in Figs. 1 and 2. The reason for this is the bio chemical activities of Sporosarcina pasteurii concrete including urea hydrolysis and calcium carbonate precipitation (bio-cementing) which decreases the porosity and increases the strength level compared to the conventional concrete. On the other hand, Rhizobium leguminosarum concrete included bacterial biopolymer clogging the spaces between the particles, act as glue holding the particles together, filling the pore spaces leading to decrease porosity and increase strength. The optimum compressive strength is found to be at 10% and 20% for Sporosarcina pasteurii and Rhizobium leguminosarum, respectively (Table 2).

Table 1. Results of compressive strength test for all mixes

Bacteria used	Bacteria,%	Average compressive strength, MPa	
		7 days	28 days
Sporosarcina pasteurii	0	15.33	18.31
	5	17.87	23.25
	10	27.17	39.66
	20	26.44	35.30
Rhizobium leguminosarum	0	19.03	24.70
	5	20.34	27.17
	10	26.44	35.45
	20	32.11	39.66

Table 2. Increase rate in compressive strength for all mixes

Bacteria used	Bacteria,%	Increase in compressive strength, %	
		7 days	28 days
Sporosarcina pasteurii	0	0	0
	5	17	27
	10	77	117
	20	72	93
Rhizobium leguminosarum	0	0	0
	5	7	10
	10	39	44
	20	69	61

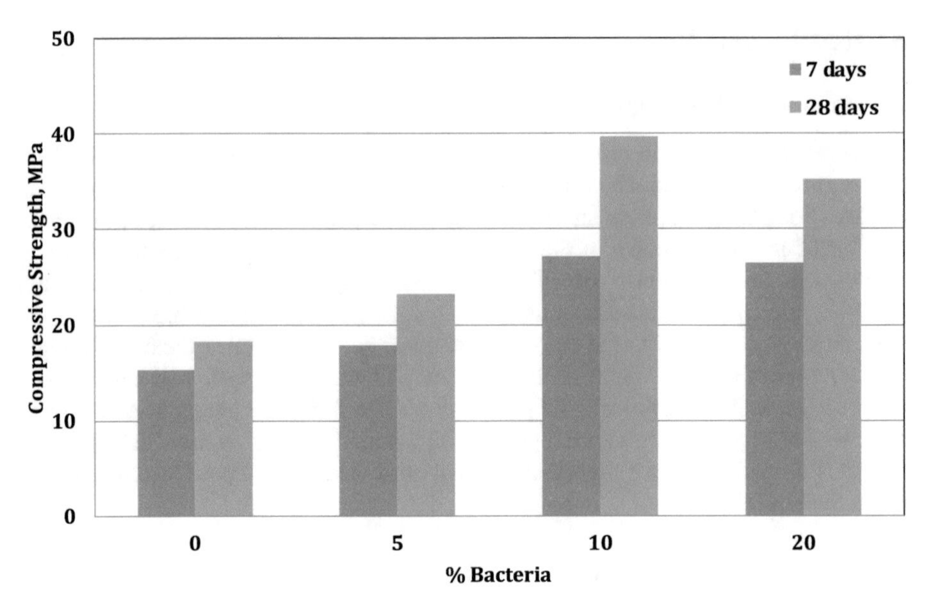

Fig. 1. Compressive strength versus Bacteria percentage for *Sporosarcina Pasteurii* after 7 and 28 days

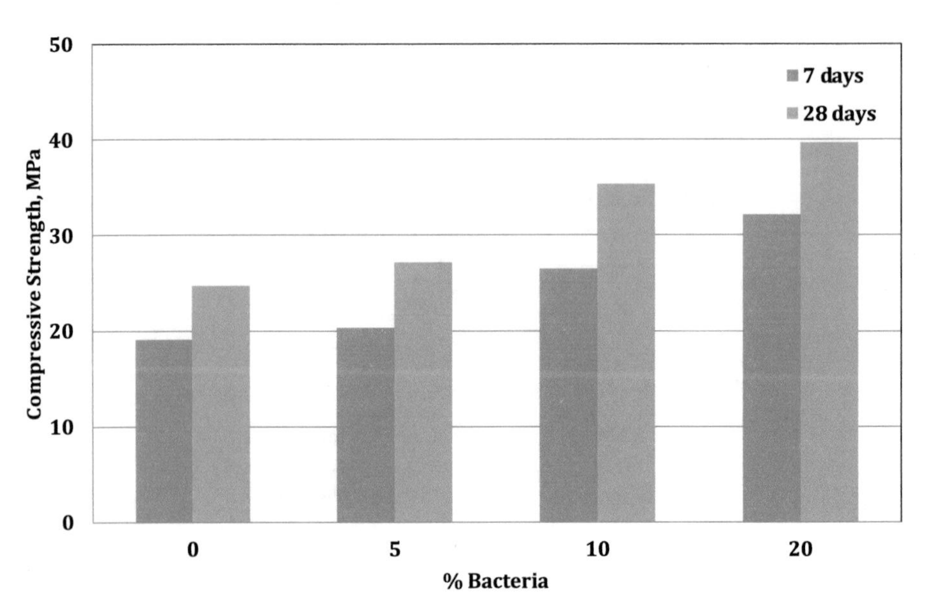

Fig. 2. Compressive strength versus Bacteria percentage for Rhizobium leguminosarum after 7 and 28 days

(2) **Split Tensile Strength**

The split tensile strength test results are tabulated in Table 3. The optimum split tensile strength is found to be at 10% and 20% for Sporosarcina pasteurii and Rhizobium leguminosarum, respectively as shown in Fig. 3. The Split tensile strength of 10% Sporosarcina pasteurii and 20% Rhizobium leguminosarum is higher than conventional concrete by 98% and 73%, respectively, this is very desired property in concrete.

Table 3. The results of split tensile strength test for all mixes

Bacteria used	Bacteria,%	Average split tensile strength after 28 days, MPa	Increase in split tensile Strength, %
Sporosarcina pasteurii	0	1.14	0
	5	1.75	54
	10	2.26	98
	20	2.03	78
Rhizobium leguminosarum	0	1.76	0
	5	2.02	15
	10	2.52	43
	20	3.05	73

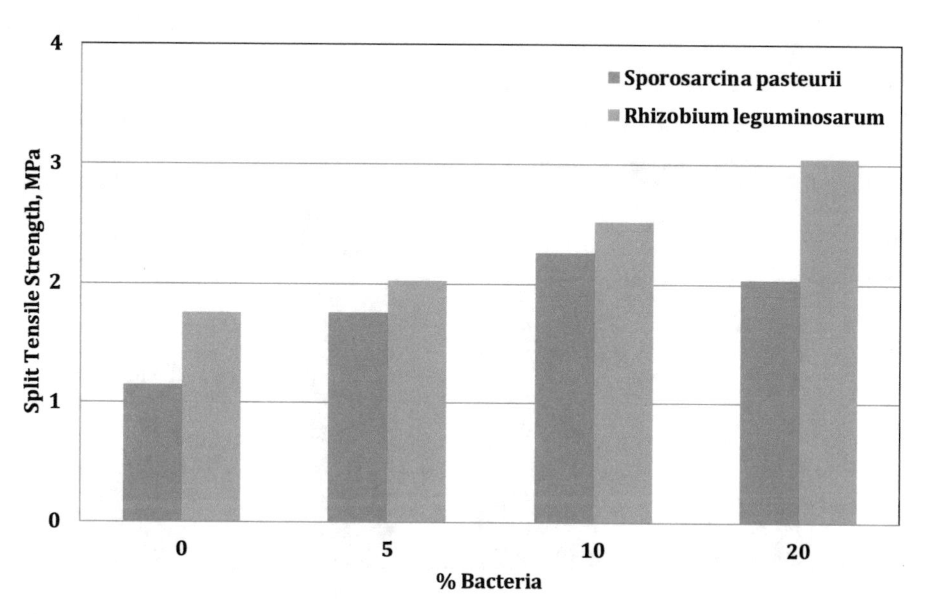

Fig. 3. Split tensile strength versus Bacteria percentage for Sporosarcina pasteurii and Rhizobium leguminosarum after 28 days

(3) **Flexural Strength**

The flexural strength test results are tabulated in Table 4. The flexural strength of Rhizobium leguminosarum concrete and Sporosarcina pasteurii concrete is greater than control concrete in all the proportions. The maximum flexural strength is found to be at 10% and 20% for Sporosarcina pasteuriiand Rhizobium leguminosarum, respectively as shown in Fig. 4. The flexural strength of 10% Sporosarcina pasteurii and 20% Rhizobium leguminosarum is higher than conventional concrete by 54% and 75%, respectively.

Table 4. The results of flexural strength test for all mixes

Bacteria used	Bacteria,%	Average flexural strength after 28 days, MPa	Increase in flexural strength, %
Sporosarcina pasteurii	0	2.52	0
	5	2.89	15
	10	3.88	54
	20	3.19	27
Rhizobium leguminosarum	0	2.71	0
	5	2.94	8
	10	3.93	45
	20	4.75	75

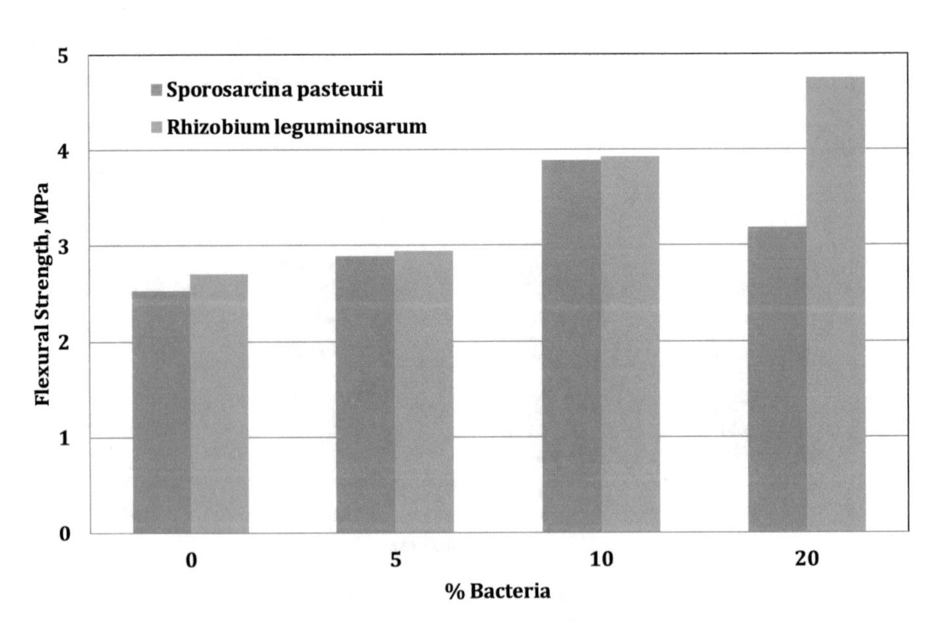

Fig. 4. Flexural strength versus Bacteria percentage for Sporosarcina pasteurii and Rhizobium leguminosarum after 28 days

(4) **Ultrasonic Pulse Velocity**

Ultrasonic pulse velocity test was carried out to know the presence of voids in the internal structure of the bacterial concrete cubes. The results so obtained after conducting the test are tabulated in Table 5. The results show that the velocity of 10% Sporosarcina pasteurii and 20% Rhizobium leguminosarum have the same highest effect on the conventional concrete by an increase of 19% (Fig. 5).

Table 5. The results of ultrasonic pulse velocity test for all mixes

Bacteria used	Bacteria,%	Ultrasonic pulse velocity after 28 days, Km/s	Increase in ultrasonic pulse velocity, %
Sporosarcina pasteurii	0	3.95	0
	5	4.41	12
	10	4.69	19
	20	4.55	15
Rhizobium leguminosarum	0	4.29	0
	5	4.55	6
	10	4.80	12
	20	5.09	19

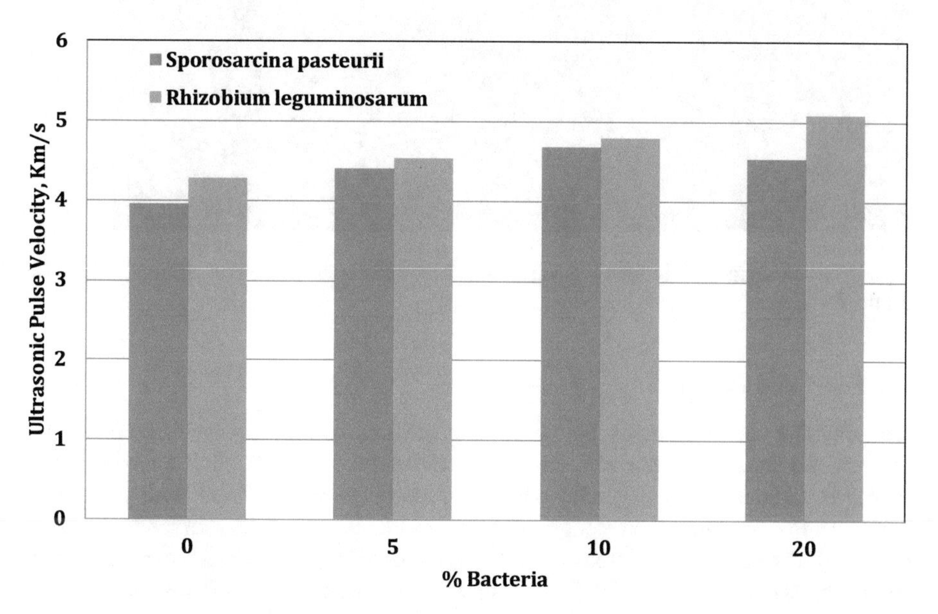

Fig. 5. Ultrasonic pulse velocity versus Bacteria percentage for Sporosarcina pasteurii and Rhizobium leguminosarum after 28 days

(5) **CT-Scanning test:**

The X-Ray Ct scanning technique was used to analyze the porosity of concrete samples with bacteria and control concrete. Figure 6 show segmented images for control concrete and bacteria concrete specimen along the depth of 15 cm. In the segmented image, the black spots represent voids between concrete particles; however, the gray color represents the concrete particles. The images were analyzed to estimate the voids (black spots) of the concrete samples, as porosity is a fraction of the voids volume over the total volume of the sample. The void spaces in the control concrete sample is greater than bacterial concrete. The least void on 10% *Sporosarcina pasteurii* bacteria concrete by produces the calcite.

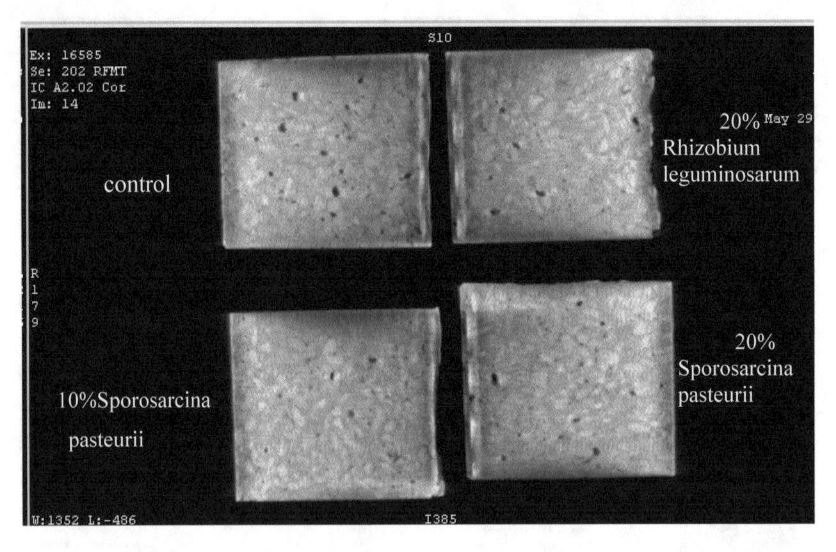

Fig. 6. Show segmented images for control concrete and bacteria concrete specimen along the depth of 15 cm.

(6) **SEM and EDX test:**

The scanning electron microscopic images of the control concrete specimen show that the particles smooth surface with large voids in between (Fig. 7). EDX analysis of this specimen indicated that the main concrete component was CaCO3 with percentage 5.61%. Show a minor percentage of other elements on control concrete listed in (Table 6). Figure 8 show decrease void between particles on bio concrete and the white color represents the calcite. EDX analysis of show increase CaCO3 on bacteria concrete with percentage 14.47%. Show a minor percentage of other elements on bio concrete listed in (Table 7).

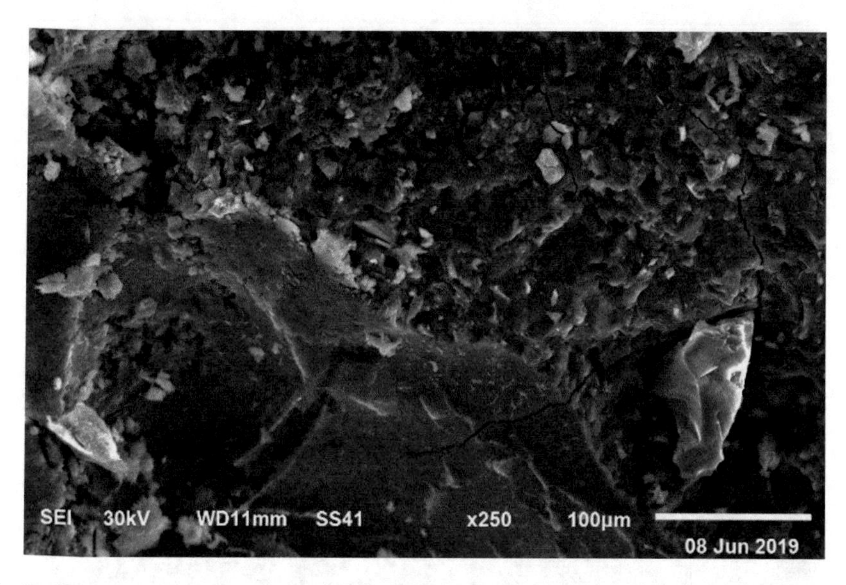

Fig. 7. Scanning electron micrographs of the control concrete at magnification powers (x250)

Table 6. The percentage of element on control concrete.

Element	Wight
CaCO3	5.6
SiO2	32.7
Al2O3	1.0
Ca	2.4
Fe	0.6

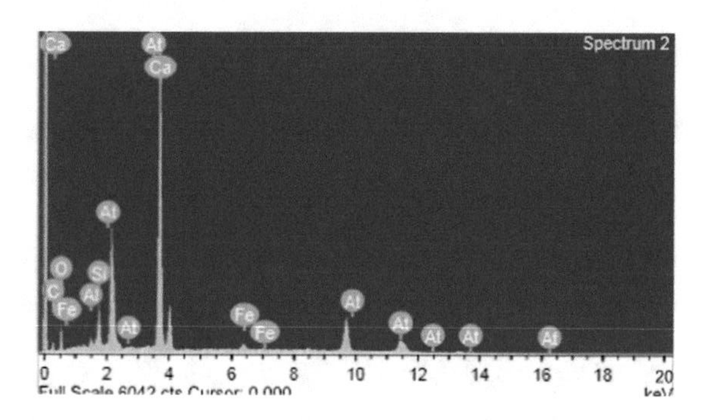

Table 7. The percentage of element on bio concrete.

Element	Wight
CaCO3	14.4
SiO2	53.1
MgO	0.3
Al2O3	2.0
SiO2	6.2
Ca	23.1
Fe	0.6

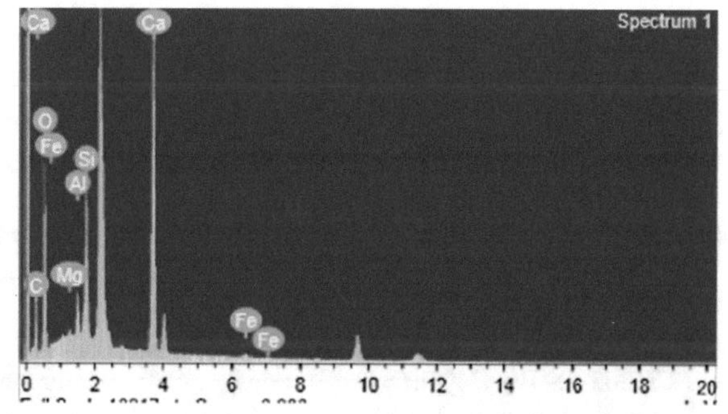

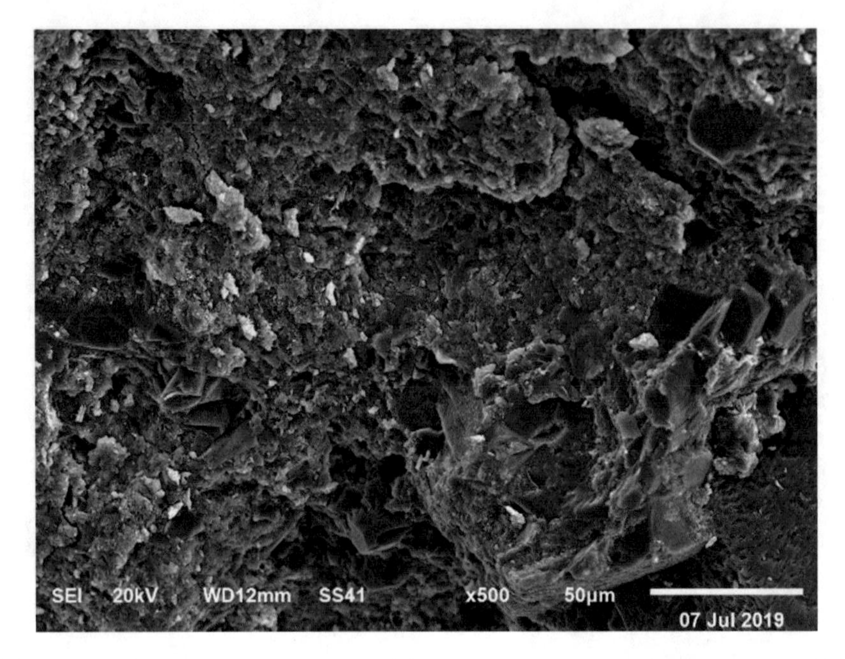

Fig. 8. Scanning electron micrographs of the bio concrete at magnification powers (x500)

6 Conclusions

Based on the results of experimental tests, the following conclusions can be drawn:

(1) Bacterial concrete can effectively increase the strength of concrete by closing the micro pores in concrete and making it denser.
(2) The 7th and 28th day strength of all mechanical properties of bacterial concrete are higher in compressive strength, split tensile strength and flexural strength, ultrasonic pulse velocity test than control concrete.
(3) The optimum percentages of bacteria are found to be 10% and 20% for Sporosarcina pasteurii and Rhizobium leguminosarum, respectively.
(4) improve the micro properties by produce the calcite and decreases void on concrete.

References

1. Pacheco-Torgal, F., Labrincha, J.A.: Biotech cementitious materials: some aspects of an innovative approach for concrete with enhanced durability. Constr. Build. Mater. **40**, 1136–1141 (2013)
2. Kumar, J.B., Prabhakara, R., Pushpa, H.: Bio-mineralisation of calcium carbonate by different bacterial strains and their application in concrete crack remediation. Int. J. Adv. Eng. Technol. **6**, 202–213 (2013)
3. ES 2421/(2005): Egyptian standard specification - Cement-physical and mechanical tests
4. ECP 203-(2017): Egyptian code for design and construction of reinforced concrete Structures, Ministry of housing, Utilities and Urban Communities, Cairo, Egypt

Author Index

M. Shehata et al. (Eds.): GeoMEast 2019, SUCI, p. 101, 2020.
https://doi.org/10.1007/978-3-030-34249-4

Printed in the United States
By Bookmasters